WHY CAR

WHY CARE?

Children's Rights and Child Poverty

W. Vandenhole, J. Vranken and
K. De Boyser (eds.)

Antwerp – Oxford – Portland

Distribution for the UK:
Hart Publishing Ltd.
16C Worcester Place
Oxford OX1 2JW
UK
Tel.: +44 1865 51 75 30
Email: mail@hartpub.co.uk

Distribution for the USA and Canada:
International Specialized Book Services
920 NE 58th Ave. Suite 300
Portland, OR 97213
USA
Tel.: +1 800 944 6190 (toll free)
Tel.: +1 503 287 3093
Email: info@isbs.com

Distribution for Austria:
Neuer Wissenschaftlicher Verlag
Argentinierstraße 42/6
1040 Wien
Austria
Tel.: +43 1 535 61 03 24
Email: office@nwv.at

Distribution for other countries:
Intersentia Publishers
Groenstraat 31
2640 Mortsel
Belgium
Tel.: +32 3 680 15 50
Email: mail@intersentia.be

Why Care? Children's Rights and Child Poverty
W. Vandenhole, J. Vranken and K. De Boyser (eds.)

© 2010 Intersentia
Antwerp – Oxford – Portland
www.intersentia.com

Cover illustration: Danny Juchtmans

ISBN 978-94-000-0025-4
D/2010/7849/84
NUR 828

No part of this book may be reproduced in any form, by print, photoprint, microfilm or any other means, without written permission from the publisher.

FOREWORD

At the dawn of what could be a new era in EU policy-making, Europe is facing a range of economic, financial and ecological challenges that threaten its progress towards a more cohesive, equal and inclusive future. Although many European leaders remain convinced that fighting poverty and social exclusion should remain a crucial item on the political agenda, that the Lisbon strategy has been quite unsuccessful in moving Europe towards achieving its ambitious goal "to make a decisive impact on poverty by 2010" and that social concerns have gradually disappeared from the Lisbon agenda should fill all Europeans with great concern.

But there is still hope. The new time horizon of Europe 2020 has inspired European leaders also to put forward a new strategy to tackle poverty. At the moment of publication of this book, its ratification is still underway and some dark clouds loom over the fixing of specific targets for poverty reduction. But although the sound of the qualification "new" rightly invokes considerable scepticism, the hard lessons learned from the Lisbon process will – hopefully – urge and inspire policy-makers to act differently and more effectively.

One of the silver linings is that tackling childhood poverty seems to have become one of the major themes in the EU's social policy-making process. Belgium has the honour of taking the lead in this matter during its Presidency of the EU in the second half of 2010, and is determined to make a success of the European Year against Poverty and Social Exclusion. Reaffirming the political determination to fight poverty deserves to be and to remain at the top of the European agenda. During our Presidency, we are putting forward child poverty as one of three selected core challenges for the EU's fight against poverty.

This book is, however, concerned not only with childhood poverty but also with the necessary complement of children's rights. This is particularly relevant in the EU context as children's rights are moving only slowly into the EU debate. It deserves to be at the heart of the childhood poverty policy discussion, as poverty today violates these rights for one in five children in the EU. We therefore wholeheartedly welcome this book as a practical academic exercise in translating the results of research on children's rights and childhood poverty into policy-inspiring readings, and in finding the crossroads between the several disciplines concerned with childhood poverty.

Philippe Courard
Secretary of State for social integration and anti-poverty policy, Belgium

CONTENTS

Foreword v

1. Introduction: why a book on children's rights and childhood poverty?
Jan VRANKEN, Wouter VANDENHOLE and Katrien DE BOYSER 1

1. Human rights and socio-economic research on poverty 1
2. The EU, childhood poverty and children's rights. 3
   2.1. Attention for childhood poverty… 3
   2.2. … and for children's rights 4
   2.3. 2010: a new momentum for addressing childhood poverty? 5
3. The encounter between children's rights and poverty approaches 6
4. Overview of chapters. 9
References 14

2. A children's rights perspective on poverty
Wouter VANDENHOLE 15

Introduction 15
1. Children's rights. 16
   1.1. Children's rights as human rights 16
   1.2. Images of children and childhood in the CRC 19
2. Children's rights and poverty 20
   2.1. Human rights and poverty – Is poverty a violation of human
        rights? 20
        2.1.1. A right to protection against poverty 21
        2.1.2. Human rights-based approaches to poverty 23
   2.2. The CRC and the Committee on the Rights of the Child and poverty 26
   2.3. Are children's rights biased against the poor? 28
3. Some tentative conclusions with potential policy relevance. 29
References 30

Contents

3. Child poverty, social exclusion and children's rights: a view from the sociology of childhood
   Virginia MORROW 33

Introduction 33
1. The sociology of childhood 34
2. Child poverty 36
3. The UN CRC and child poverty 38
4. Social exclusion and human capabilities approaches: linking to the UN CRC 40
5. Social exclusion (and social capital) 44
6. Children's participation 48
7. Conclusion 50
Acknowledgements 51
References 51

4. Child poverty, children's rights and participation: a perspective from social work
   Rudi ROOSE, Griet ROETS, Didier REYNAERT and Maria BOUVERNE-DE BIE 57

Introduction 57
1. In the name of participation 58
2. Participation as an instrument 60
3. Participation as a point of departure 63
4. Conclusion 65
References 66

5. Children in public care in England: well-being, poverty and rights
   Nina BIEHAL and Gwyther REES 71

Introduction 71
1. Child well-being 72
    1.1. The well-being of children in public care 74
2. Child poverty and well-being 75
    2.1. Well-being and poverty for children in public care 76
3. The rights of children in and on the edge of public care 79
    3.1. Children's rights in England 79
    3.2. English child care policy: balancing the rights of children and parents 81
    3.3. Protection 83
    3.4. Non-discrimination 84
    3.5. Participation in decision-making 85

4. Conclusion: well-being, poverty and rights. 86
References 87

6. 'Poverty, like beauty, lies in the eye of the beholder'?
   Jan VRANKEN 91

Introduction 91
1. What UNICEF says. 92
2. Different perspectives 94
3. Different poverty models in the social sciences 95
   3.1. Economic welfare: from a one-dimensional to a multidimensional
        conceptualisation 95
   3.2. The capabilities approach 99
   3.3. Poverty as a form of social exclusion 100
4. Is there a culture of poverty? 102
5. From a static to a dynamic (life course) perspective 104
6. The importance of networks 106
7. Between the social and the spatial: does social cohesion imply social
   exclusion? 107
8. Is this a relevant framework for understanding child poverty? 108
References 110

7. Child poverty – What's in a word?
   Keetie ROELEN 113

Introduction 113
1. The concept of child poverty – theoretical considerations 115
   1.1. Universal versus context-specific 116
   1.2. Well-being versus well-becoming 117
   1.3. Age matters 119
2. The concept of child poverty – empirical implications 120
   2.1. Universal versus context-specific 120
   2.2. Well-being versus well-becoming 122
   2.3. Age matters 124
3. Conclusion 126
References 128

8. Escaping poverty with your children: the role of labour market
   activation, education, and social capital investments
   Ingrid SCHOCKAERT and Ides NICAISE 131

Introduction 131
1. Theoretical perspective 132

2. Data 135
3. Method 137
   3.1. The model estimation...................................... 137
   3.2. The simulation model...................................... 138
   3.3. Adding children to the picture 139
4. Results 139
   4.1. Poverty outflow, education, social capital and employment 139
   4.2. The potential impact of policy measures................... 141
5. Discussion.. 141
   5.1. Poverty dynamics, education, social capital and employment 142
   5.2. The impact of general anti-poverty measures on the persistence
        of child poverty 142
6. Conclusion 143
Annex 1 144
Annex 2 145
Annex 3 147
References 150

9. **Early childhood poverty in the EU: making a case for action**
   Katrien DE BOYSER 153

1. Childhood poverty as a social research and policy concern...... 153
2. Why focus on poverty in the earliest life stage? 154
   2.1. Is early deprivation too often a missing element in the social
        mobility debate? 154
   2.2. Poverty, early life health and health throughout the life-course..... 155
   2.3. On early childhood poverty and development: is poverty a
        brain drain? 156
3. Early childhood poverty and deprivation in the European Union 157
   3.1. Economic deprivation and lack of durables 159
   3.2. Nutritional deprivation 160
   3.3. Housing deprivation.. 161
   3.4. Multiple deprivation....................................... 162
4. The European Union and early childhood policies 163
5. To conclude: a strong case for action 164
References 165

10. **A school in the neighbourhood, a neighbourhood in the school**
    Isabelle PANNECOUCKE 169

Introduction 169
1. The contextualised school 169
   1.1. Experiencing school life................................... 170

	1.2. A meeting place	170
	1.3. A place to play	172
	1.4. A place to learn	172
2.	Choosing a school	173
3.	Lessons for policy makers	178
	References	180

11. Like a child's game: a policy configuration approach to child poverty
Danielle DIERCKX 183

Introduction		183
1.	The policy configuration approach	183
2.	Child poverty as a policy theme	184
	2.1. A bunch of definitions	184
	2.2. The lie of good intentions	185
	2.3. Agenda setting	186
3.	A holistic approach guaranteed by governance structures	188
4.	Who plays a decisive role?	191
	4.1. A life experience approach	191
	4.2. An academic evidence-based approach	192
5.	Conclusion	192
	References	193

About the authors 195

1. INTRODUCTION: WHY A BOOK ON CHILDREN'S RIGHTS AND CHILDHOOD POVERTY?

Jan Vranken, Wouter Vandenhole and Katrien De Boyser

1. HUMAN RIGHTS AND SOCIO-ECONOMIC RESEARCH ON POVERTY

Research on *poverty* has developed significantly in the past two decades, with important contributions to the debate both in terms of conceptualising and explaining poverty and in terms of developing methodological tools for its analysis. The specificity of poverty among children has also received more attention in socio-economic research and the importance of a specific policy focus on children is now widely recognised.

Previously, children were not a focus of analysis in poverty research and little effort went into addressing specific problems of assessing poverty among children. What was missing in many studies was the understanding that poverty affects children in different ways than adults, and that it can have different consequences for them. In fact, in many cases child poverty was treated as a subset of overall or adult poverty, using the same methods and techniques.

Then and more recently, the potential role of a *human rights approach* to socio-economic research on poverty has not been fully explored, at least not in terms of how human rights can influence research questions, the focus and structure of analysis and the identification of policy implications. Despite increasing cross-referencing between research on poverty and research on human rights, they seem mainly to have followed two different and even separate tracks. Simply referring to human rights, and human rights instruments, in poverty research does not mean using a human rights approach: this is not enough to have human rights principles reflected in socio-economic research. Conversely, human rights research reflects difficulties in assuming some of the methodological complexities of socio-economic research.

UNICEF (2006) identifies three key factors that support a specific focus on child poverty. These are time, agency and the role of public services.

The importance of *time* in the analysis of child poverty stems from the dual views on children as both "being" (having their own importance in current time) and "becoming" (developing into adults). This second focus has become important since longitudinal data became available, through which the dynamic aspect of poverty can be studied. In the short term, children can be deprived relative to their peers but in the longer term, poverty and deprivation can impact on child development, with long-term consequences also in adulthood. The "becoming" view is more concerned with the loss of human capital for society.

A clear rationale for focusing on child poverty derives from the fact that children are rapidly evolving and that their evolving capacities need to be assessed and protected in their own right. Also within this time perspective, poverty is likely to mean different things, both subjectively and objectively, to children of different ages. This has consequences for the measurement of poverty among children: with whom should they be compared – everybody, just other children, or an even narrower age sub-group of children?

Agency is a second key factor in the UNICEF study; it refers to one of the key forces that drive policies to reduce child poverty, the dependency of the child. Children are said to have little impact on their own situation. It is through the agency of parents in combination with exogenous forces that lie outside the family's control, and to a lesser extent the state, that children's situations are shaped. This is recognised in the Convention on the Rights of the Child (Article 27(2)), which provides that "[t]he parent(s) or others responsible for the child have the primary responsibility to secure, within their abilities and financial capacities, the conditions of living necessary for the child's development". In examining and explaining the well-being of children, the family context and wider societal conditions cannot be ignored.

Agency is also important on the macro level. When considering the evolving capacities of the child, the role of agency is important in terms of policies to support children. Such policies can be directed at children themselves or at parents, either in the form of direct transfers or in terms of incentives such as childcare subsidies to encourage mothers to take up paid employment. Any policies that aim at reducing child poverty cannot ignore the particular triangular relationship that exists between children, their parents and the state. Children, then, embody the "public good" and therefore *public services* should be concerned with their situation. Successful states need healthy and educated children and in most countries the government has committed itself to ensuring children's access to at least a basic level of health and educational services.

1. Introduction: why a book on children's rights and childhood poverty?

Children's dependence on public services suggests that family income and consumption, not a perfect measure of well-being in any context, may be somewhat less revealing in the case of child poverty if not measured in conjunction with indicators of access to services. Mehrotra and Delamonica (2002) highlight the fact that public service provision is less effective when there is poverty at home, and also that in a context of poor public service provision, children profit less from improvements in the general poverty situation.

2. THE EU, CHILDHOOD POVERTY AND CHILDREN'S RIGHTS

The issues of childhood poverty and children's rights have received increased attention from policy-makers at the EU level and across European countries over the past decade.

2.1. ATTENTION FOR CHILDHOOD POVERTY...

The need to prioritise *child poverty* has now been emphasised by three successive EU Council Summits, and since the adoption of the Lisbon strategy ever more countries have specifically targeted their efforts at combating poverty in childhood in their national inclusion strategies. Many member states which made the issue a priority in 2006 in response to the European Council's call for decisive action have mainstreamed child poverty in areas such as minimum income and wages, reconciliation of work and family life, and family-friendly services. Drawing on the improved evidence base, 22 member states have set targets in relation to child poverty, 16 of them using EU-agreed indicators. A few have also set intermediate targets for their specific challenges (jobless households, families most at risk, intensity of poverty, childcare). (European Council, 2009: 6). Childhood poverty was a key focus theme for the year 2007 in the context of the Open Method of Coordination of social protection and social inclusion, as it is again in 2010, the European Year for Combating Poverty and Social Exclusion.

Why this increased attention for children in poverty? One reason is that children are more likely to experience poverty than the population as a whole – 19% compared to 16% – and this situation has not improved since 2000 (European Council, 2009: 6). Moreover, poverty in childhood very often reduces any hope of escaping poverty in adulthood, which may have significant long-term effects not only for the individuals concerned but also for society and the economy at large. In current policy debates, these findings lead to concerns with low education levels, the loss of human capital available for the (knowledge) economy, and with the financial burden on the (future) social budget of the state. Last but not least, concern with child poverty is rooted in the very old perception of children as innocent and thus as "deserving poor" – meaning that they cannot be blamed for their poverty and therefore deserve compassion, charity and assistance.

Does the increased attention given to childhood poverty also lead to more efficient strategies and thereby less childhood poverty? As for poverty in general, the EU has little means at its disposition for developing a coherent and strong strategy for combating child poverty; what exists is of the "soft law" type. The Social Open Method of Coordination was established in 2000 to help the EU achieve its objective of "making a decisive impact on the eradication of poverty". It binds member states to common objectives, common indicators and a common reporting process, and aims at facilitating progress through peer reviews, stakeholder involvement and research. The question is whether the recent "enlargements" will favourably impact on the position of child poverty in the political agenda. The first enlargement is about the inclusion of the former NAP/Incl in National Strategy Reports on Social Protection and Social Inclusion, which contain not only "Social Integration" (as Inclusion is now called) as an item, but also the "harder" items "pensions" and "health and long-term care". It remains to be seen whether social integration will continue to receive strong and sustained attention.

The other "enlargement" refers of course to the rapid increase in the number of EU member states. News stories of extreme child abuse in some of these newcomers may lead to an overemphasis on basic child protection and a neglect of the more "sophisticated" aspects of recently developed policies on children, such as those on child poverty.

2.2. ... AND FOR CHILDREN'S RIGHTS

Children's rights have also received more attention from the EU over the past couple of years. By incorporating an explicit reference to children's rights into the Treaty on European Union, which entered into force on 1 December 2009, a clear legal basis has been created for an EU children's rights policy. Policy documents on children's rights in EU action have been produced, such as the Commission Communication *Towards an EU Strategy on the Rights of the Child* (European Commission, 2006), the *EU Guidelines for the Promotion and Protection of the Rights of the Child* (European Council, 2007), and the Commission Communication *A Special Place for Children in EU External Action* (European Commission, 2008).

1. Introduction: why a book on children's rights and childhood poverty?

While this rather recent interest in children's rights is a most welcome development, and may offer additional leverage for a solid and lasting policy on child poverty, there is no reason for naïve optimism. First of all, most EU documents produced so far tend to focus primarily if not exclusively on the EU's external *actions* concerning so-called third countries, rather than addressing children's rights issues within the EU. Secondly, the children's rights agenda tends to be subordinated to or instrumentalised for concerns with the ageing of Europe, an economic agenda in the fight against child poverty, a security agenda in migration policies, or a punitive approach in the fight against human trafficking. Thirdly, there seems to be little understanding of a consistent approach of children's rights, so that often mere lip service is paid by introducing rather superficial references (Stalford & Drywood, 2009: 155, 161). Moreover, a concern with the need to protect children seems to prevail over recognition of their agency and the need for their involvement and participation. It remains to be seen whether the long overdue EU strategy on the rights of the child will be able to remedy these shortcomings. Only then might children's rights provide additional leverage for an effective EU strategy to address child poverty, although the more fundamental questions about the relationship between children's rights and poverty should always be kept in mind (see below).

2.3. 2010: A NEW MOMENTUM FOR ADDRESSING CHILDHOOD POVERTY?

What was already clear to many during the Mid-Term Review of the Lisbon process in 2005, that it was very unlikely that a significant reduction in poverty would be realised by 2010, proved to be a correct presumption. Even though the Lisbon strategy may have given an impulse to the development of poverty policies at national level, the results of those policies remained far below expectations, and the whole process was generally evaluated as a failure. As the Lisbon Agenda ends on 31 December 2010, a new approach is needed that also takes the new EU constellation of 27 – very diverse – member states into account.

In March 2010, the Commission issued a communication on a new ten-year strategy called "Europe 2020, a strategy for smart, sustainable and inclusive growth". Following this communication and the discussions held in the European Council on 25 and 26 March 2010, the Council reached an agreement on the new strategy, which will be formally adopted in June. It is expected that its general lines will remain intact.

One of the main objectives is a clear poverty reduction target, set at reducing the number of people living below the poverty line by 25%. Other related targets are a reduction in the share of early school leavers to below 10% (currently 15%) and raising the employment rate of the population aged 20–64 from the current 69% to 75%. The main targets are also translated into so-called flagship initiatives: for the reduction of poverty, the European Platform Against Poverty must "ensure social and territorial cohesion such that the benefits of growth and jobs are widely shared and people experiencing poverty and social exclusion are enabled to live in dignity and take an active part in society" (EC, 2010).

It is hard for many stakeholders not to be sceptical about this new agenda, which might suffer from the same lack of political will at national level as was the case for the Lisbon principles. Both were, also according to the Commission, important factors in the failure of the former strategy. And indeed, national governments have questioned the EU's legal right to set targets on education and poverty; they could only agree broadly on the importance of improving education and tackling social exclusion. Quantifiable targets are not now expected until June and any explicit reference to cutting poverty by 25% has been dropped in the meantime, as have targets on reducing the numbers of early school leavers and increasing participation in tertiary education.

From a more scientific point of view, there is the quite fundamental issue of the adequacy of the indicators underpinning the process of monitoring national and EU policy progress on poverty targets. With the enlargement of the EU to 27 member states, which differ greatly in terms of living standards, the main indicator for measuring poverty – the at-risk-of-poverty rate – has come under fire. As it stands, this relative measure no longer adequately brings the relative poverty position of member states into perspective, as it measures relative poverty or income inequality within one state. This results in similar at-risk-of-poverty rates for the poorest and the richer member states, while the whole context, such as the welfare regime and the presence of other public services, may greatly differ between them.

3. THE ENCOUNTER BETWEEN CHILDREN'S RIGHTS AND POVERTY APPROACHES

As stated earlier, childhood poverty has attracted attention from different social science disciplines – such as psychology, educational science, economics, law and sociology – and more recently also from the life sciences (biology, neuroscience). Gradually, child poverty scholars have developed an interest in a (Anglo-Saxon) rights-based perspective, as it was first used during the working period of the "European Observatory on National Policies to Combat Social Exclusion" (see Room, 1992 and following). In a rights-based approach, social exclusion in general is defined in terms of being denied or not realising basic social rights.

1. Introduction: why a book on children's rights and childhood poverty?

Even though in policy documents this structural rights-based approach is often taken as the framework, the policies proposed often have too little impact or are too fragmented to really live up to this general idea.

On the other hand, child poverty has generated some interest amongst part of the children's rights community – activists and scholars alike. Clearly, as several authors point out, children's rights and child poverty studies have more often than not followed parallel tracks. So how do both perspectives look at child poverty and at each other, and what can be gained from bringing them together?

From a children's rights perspective, activists tend to qualify child poverty as a violation of children's rights. Child poverty should therefore be ended immediately. Children's rights are also believed to offer some guidance and even tools for eradicating child poverty. Firstly, children's rights increase the strength of the argument, partly because they hold a strong moral claim, partly because they invoke legal norms that have been accepted by an overwhelming majority of states, currently 193. A children's rights perspective also represents a paradigmatic shift from charity to rights, with the concomitant principles of participation, accountability, non-discrimination, empowerment and linkage with human rights. Secondly, it allows a superseding of ideological positions and clashes, because human rights are assumed to have the support of a majority of the population – excluding some extreme fractions. Thirdly, inserting human rights-based approach (HRBA) principles into policies and programmes on child poverty also narrows policy discretion, in that a minimum threshold is established; process requirements such as participation and prioritisation of the poor, and of poor children in particular, are introduced; and the best interests of the child are to be taken into account.

While a reference to children's rights may therefore seem very attractive and powerful, and undoubtedly holds some potential that should be neither underestimated nor downplayed, at least part of the community of children's rights scholars tends to be more cautious about the leverage and quick fix capacity of children's rights. First of all, the relationship between child poverty and children's rights *law* is not that clear-cut. Poverty is explicitly mentioned neither in the 1989 Convention on the Rights of the Child (CRC) nor in any of the other core international human rights treaties. No right to be free from poverty or to be protected against poverty has been recognised in the CRC. This should not come as a surprise. The duty-holder under human rights law, including the CRC, is primarily if not exclusively the state. Few if any states would be tempted to impose on themselves a legal obligation to eradicate child poverty, not even as an obligation of means rather than of result. So technically speaking, poverty can only be argued to constitute a *violation* of children's rights if the *state* has failed to abide by its obligations.

Human rights law, to which the CRC belongs, also tends to take an individualistic rather than a collective or structural approach. The question can be raised therefore whether and to what extent the lack of a clear reference right in the CRC for addressing poverty, in combination with the statist and individualistic bias of human rights law, leads to an inherent tendency within children's rights law to approach child poverty in a fragmented manner, in which certain issues are singled out (social assistance or housing or medical assistance) but not comprehensively taken up. Moreover, root causes may remain unaddressed. The underlying assumption in children's rights law of the autonomy of the individual may reinforce this risk of individuating and decontextualising poverty.

Thirdly, the conceptual advantages of a human rights-based approach do not materialise automatically into practice as in a human rights-based approach the state's responsibility with regard to poverty is a subsidiary one, which is, moreover, centred on material conditions and access to social services. One may wonder whether this approach corresponds to current understandings of the nature of poverty.

Fourthly, there is an ongoing debate on fundamental questions such as whether and to what extent poverty impinges on children's ability not only to resist power, but also to change power relations and the concomitant social exclusion. If poverty does so, the empowering potential of a children's rights approach is seriously limited, for such an approach presupposes rather than "creates" empowered, participating individuals who are able and willing to invoke their rights, if necessary before a court of law, in order to hold the state accountable. Caution may be warranted so as not to place too much responsibility on those who may not have the power to change their circumstances fundamentally.

Finally, an instrumental approach to children's participation – and by extension to children's rights more generally – in addressing child poverty tends to dominate. This instrumental approach is part of a broader trend towards considering children in poverty as the victims of poor parenting, thereby emphasising the victimhood of the children and the individual responsibility of the parents, rather than identifying and challenging the structural causes of poverty.

So children's rights may have something to offer child poverty studies and poverty policies: they possibly provide some common ground for the different perspectives on the causes of poverty, given their broad support in society. They also introduce some process requirements, in particular the participation of the poor, also in the way research is conducted. So the perspectives of children living in poverty should feed into the study of child poverty, while keeping in mind the ambiguity of many participation practices. At the same time, children's rights

1. Introduction: why a book on children's rights and childhood poverty?

may gain from an encounter with child poverty studies, not least in grasping the complexity of child poverty and in making a realistic assessment of its potential.

4. OVERVIEW OF CHAPTERS

The book is divided into two parts. After this general introduction to the book, children's rights perspectives are offered. In the second part, child poverty is approached from the perspective of poverty studies. In this section, we provide a short overview of the content of the chapters that follow.

In recent years, writes Wouter Vandenhole, the human rights perspective has gained considerable ground, including in matters of (child) poverty. This is an important evolution, for several reasons. It increases the strength of the argument and it allows ideological positions and clashes to be superseded. Indeed, rights are assumed to have the support of a majority of the population and may thus provide some common ground to the different perspectives on the causes of poverty.

Children's rights strike an even more sympathetic chord. The very vulnerable situation of children makes them the prototype of the "deserving poor" and thus children's rights can provide an important lever for combating the poverty of children and possibly also that of their parents, and even more broadly for emancipation and social struggle. However, in such an approach children's rights are mainly associated with vulnerability and protection, and much less with agency and participation. Moreover, conceptual clarity on the relationship between human rights, including children's rights, and poverty is still missing. Finally, the limitations and pitfalls of a rights-based approach should not be ignored in allowing for an informed and strategic mobilisation of the children's rights framework.

In his overview chapter, Wouter Vandenhole addresses the issue of children's rights and child poverty primarily from an analytical perspective, i.e. how children's rights (instruments) approach child poverty. Such an analytical approach is to be distinguished from an advocacy perspective, in which children's rights are invoked as an instrument of mobilisation and political pressure. The normative framework of children's rights is succinctly presented as part of the broader human rights framework. He then examines the dominant understandings of how children's rights and poverty are related. In his tentative conclusions, he also pays attention to the policy relevance of some of the findings.

The next two chapters focus on particular approaches to, or aspects of, a children's rights approach. Virginia Morrow starts from the fact that discussions and debates on child poverty and children's rights rarely seem to connect, and pass each other like trains on parallel tracks. In spite of in her view well-established arguments that poverty and social exclusion breach human rights, the links between *children's* rights as enshrined in the UN Convention on the Rights of the Child and *child* poverty are not so clearly articulated in research. This is particularly the case in the UK. When child poverty specialists do attempt to include a discussion of the UN CRC in their work, they tend to do so in a partial way, selecting various articles of the Convention to fit their argument, and there is a danger of associating children in poverty with "victimhood".

Why is there this gap, the author wonders? She does not explore the contribution to debates about child poverty or children's rights from mainstream (as opposed to radical or critical) developmental psychology and child psychiatry. Nor does this chapter focus on the very large literature on inequality, social stratification and social class that exists in the UK – again, this body of work tends not to be based on, or engage with, debates about children's rights. Instead, she attempts to explore some of the reasons through the lens of the "new" sociology of childhood, based on a brief review of the available literature and drawing upon some of her empirical research, and suggests some ways forward.

Rudi Roose, Griet Roets, Didier Reynaert and Maria Bouverne-De Bie consider it essential to turn a critical eye on underlying problem constructions and the resulting policy and practice interventions. Putting the development of anti-poverty policy agendas in a historical and international perspective illuminates the way in which policy constructs its objects of intervention. In framing child poverty as a problem that needs interventions and solutions, child poverty is made particularly politically salient under certain ideological conditions, so that it becomes fit for future intervention by policy makers and practitioners in social work and related disciplines. During recent decades, anti-poverty policies, which include child poverty, have been ostensibly undertaken in the name of participation. Nevertheless it is often not clear what participation and the right to participate really mean, partly because the participation discussion remains under-theorised. The authors argue that participation is not an *inherently* positive and unambiguous notion at all because it mirrors the ideological motives of social policy, civil society and practice. In their contribution, they relate the question of child poverty and children's rights to participation as a key concept in the debates on combating child poverty in the area of social work.

The next chapter already moves beyond a children's rights perspective and focuses also on the wellbeing of the highly vulnerable group of children and young people in a particular situation (public care) in England. However, Nina

1. Introduction: why a book on children's rights and childhood poverty?

Biehal and Gwyther Rees discuss each of these issues in the context of the wider research on these topics. This group typically lives in family foster care, but around one in ten stay in residential children's homes. Due to their very troubled backgrounds before entering care, initial well-being may be low for many of them. Their past adversities may also have repercussions for their further well-being.

Although substitute care may provide some compensation for past adversities, there are continuing concerns that public care may also compound these difficulties instead of compensating for them. Many of these children experienced poverty within their families before entering care and there is concern that they are at particularly high risk of poverty after leaving care, in late adolescence. They are therefore highly vulnerable to a range of adversities prior to, during and after the time they are in care.

The discussion of this particularly vulnerable group raises key dilemmas in relation to children's well-being and children's rights, which may be of relevance to other marginalised groups, such as disabled or asylum-seeking children, as well as to other children and young people in the wider community. These questions are also relevant for understanding what is happening in other countries. The extent to which societies ensure the wellbeing of the most marginalised groups of children may be an important indicator of how seriously they take their responsibilities towards all children. In this respect and more than ever, the famous saying of Dr Johnson is applicable: "a decent provision for the poor is the true test of civilization" (Boswell, 1791).

The discussion then turns to poverty studies. In his opening chapter on childhood poverty, Jan Vranken focuses on the impact of the perspective that is being used – sometimes, that has been chosen – on the study of poverty. He gives two reasons for this focus. First, poverty in general constitutes the context for describing, analysing and combating child poverty. Second, the importance of the "perspective" on or the "model" of poverty that structures whatever is written or said on poverty has always been underestimated, especially in policy-making and research circles. After some remarks on the UNICEF approach, he discusses – not necessarily in this order – different perspectives on poverty, different poverty models in the social sciences, the role of the "culture of poverty" and of networks, and the importance of a dynamic (life course) perspective and of the interaction between the social and the spatial (does social cohesion imply social exclusion?). He ends with answers on the question whether a general framework suffices to understand child poverty – and links these to Keetie Roelen's contribution.

At the outset of her chapter, Keetie Roelen states that the development of child poverty approaches is a normative process, hinging on value judgments and theoretical considerations that lead to differing interpretations of reality. As a result, child poverty outcomes might diverge depending on the approach taken. In order to structure the discussion of issues that influence the choice of concept and the implications of this choice, she discusses a generic construction model she developed herself. Roelen addresses a selection of theoretical considerations related to the concept of child poverty that may have an impact on the current debate on child poverty measurement. Central is the disparity between monetary and multidimensional poverty measurement. She considers this tension to be different from that on the well-becoming versus well-being issue. Especially with respect to children, the author illustrates the implications of these theoretical considerations by providing empirical examples, which strongly indicate that an understanding of the conceptual construct of child poverty is important beyond the theoretical context and also has practical implications.

Ingrid Schockaert and Ides Nicaise analyse the dynamic concurrence of general and child poverty, and the effect on child poverty of factors that are known to influence general poverty such as employment, educational attainment and social participation. They start by presuming that factors affecting general poverty (household level) will also affect the living conditions of children, but not necessarily to the same extent. They demonstrate the potential impact of general policy measures such as labour market activation, prevention of school dropping out and social capital enhancement on the dynamics of child poverty by using an ex-post micro simulation model. They conclude their chapter by providing policy recommendations.

In her chapter, Katrien De Boyser focuses on early childhood poverty in the EU and makes a case for policy action for this age group. First, she gives an overview of the current situation of early childhood deprivation in the EU-27, taking a multidimensional perspective. This analysis immediately bumps into the measurement issues mentioned earlier, where it becomes clear that a multidimensional approach can illuminate flaws in other indicators (in this case, the at-risk-of-poverty rate) and provide a better overall base for international comparison. Secondly, the author takes us through a selection of international literature, demonstrating possible short- and long-term effects of childhood poverty in the earliest life stage. Increasing evidence is becoming available on the importance of the earliest life stage in establishing the crucial foundations for future achievement and social mobility. Finally, she looks into policy recommendations which may be relevant to EU policies on early childhood poverty and its outcomes.

1. Introduction: why a book on children's rights and childhood poverty?

The empirical basis of Isabelle Pannecoucke's contribution is even more specific: research conducted amongst children living in three suburbs in Antwerp, Belgium. However, as for the preceding chapter she gained much more general insights, in her case on the (lack of) interaction between schools and their neighbourhood, viewed through the eyes of children. Despite developments in and changed views on education and learning, school life still takes place mainly within the school walls, behind a gate keeping the outside world at a safe distance. The author starts out by demonstrating that this kind of separation is not absolute, as the school environment is not a "neutral" given. Because of this interdependence between neighbourhood and school it is important to pay attention to the spatial context and to the way children experience school life and their neighbourhood. Experience of school life is related not only to the real situation at school, but also to the specific family situation, the social and cultural context in which children live and the location of the school in the broader societal context. The feelings evoked by school life are illustrated by descriptions of the subjective attitudes of children towards their schools.

This book ends by offering a framework for the analysis of child poverty policies, based on the policy configuration approach. Danielle Dierckx distinguishes three main components of this approach: the policy theme, the policy organisation and the policy style. The policy theme of child poverty, given its multidimensional character, is assumed to put a lot of pressure on the government's (internal) organisation. Dierckx suggests a scheme to operationalise a policy organisation that guarantees coordination. Finally, the policy style with regard to child poverty, adopting a rights-based approach, should be participatory. Participation is considered as a way to generate information that informs policies and policymaking. Dierckx explores both the information flows from children experiencing poverty (the life experience approach) and from academics (the evidence-based approach). While in principle the perspective of children may enrich policies, much depends on how their participation is defined.

She concludes by pointing out some important policy guidelines. First, in order to strengthen the policy discourse a broad consensus amongst governmental and non-governmental actors involved with children and/or poverty matters should be realised. Secondly, governments need to organise themselves better to deal with the ambitions of a coordinated, integrated approach. Accountability is central to such an approach. Finally, the evidence from children's life experience may complement the classical evidence of scientific research. Governments therefore need to create opportunities and space for more interactive policy-making.

With this book, we aim to introduce several studies of child poverty and of children's rights, and to identify intersections between different theoretical approaches from childhood poverty on the one hand and the field of children's rights on the other. In addition, we hope that policy-making may benefit from this – in our view – fruitful encounter of perspectives and disciplines, by pointing towards policy issues and considerations. It goes without saying that we have not exhausted the discussion, nor have we attempted to develop a comprehensive set of policy considerations, let alone policy recommendations. But hopefully we have generated some momentum towards enabling the perspectives and disciplines to meet and enrich each other.

REFERENCES

BOSWELL, J. (1791), *The life of Samuel Johnson LL.D.*
EUROPEAN COMMISSION (2006), *Communication, Towards an EU strategy on the rights of the child*. COM(2006) 367 final.
EUROPEAN COMMISSION (2008), *Communication from the Commission to the Council, the European Parliament, the European Economic and Social Committee and the Committee of the Regions, A special place for children in EU external action*. COM(2008) 55 final.
EUROPEAN COMMISSION (2010), *Europe 2020. A European strategy for smart, sustainable and inclusive growth*. COM(2010) 2020. Brussels: EU.
EUROPEAN COUNCIL (2007), *EU guidelines for the promotion and protection of the rights of the child*.
EUROPEAN COUNCIL (2009), *Joint Report on Social Protection and Social Inclusion*.
MEHROTRA & DELAMONICA (2002), Public spending for children: an empirical note, *Journal of International Development*, (14): 8, 1105–1116.
ROOM, G. et al. (1992), *Observatory on National Policies to Combat Social Exclusion. Second Annual Report*, Brussels, Directorate General for Employment, Social Affairs and Industrial Relations, Commission of the European Communities.
STALFORD, H.K. & DRYWOOD, E. (2009), Coming of age?: Children's rights in the European Union, *Common Market Law Review*, (46): 143–172.
UNICEF (2006), Understanding child poverty in south-eastern Europe and the Commonwealth of Independent States, *Innocenti Social Monitor 2006*. Florence: UNICEF Innocenti Research Centre.

2. A CHILDREN'S RIGHTS PERSPECTIVE ON POVERTY

Wouter VANDENHOLE

INTRODUCTION

In recent years, the rights perspective has gained considerable ground in a multitude of disciplines and approaches. It has been introduced into the consideration of many societal issues, including poverty.[1] The importance of a rights perspective should not be underestimated, for several reasons. Firstly, it increases the strength of the argument, partly because it invokes legal norms that have been accepted by the state. Secondly, it allows ideological positions and clashes to be superseded because rights are assumed to have the support of a majority of the population – some extreme fractions not included. Rights may thus provide some common ground for the different perspectives on the causes of poverty that are present in all sections of society (see Vranken chapter 6).

As to child poverty, children's rights strike an even more sympathetic chord. The broadly shared assumption that children are in a very vulnerable situation makes them the prototype of "deserving poor", i.e. those that deserve to be assisted and protected. Children's rights can provide an important lever for combating the poverty of children and possibly also that of their parents, and even more broadly for emancipation and social struggle. However, a price may have to be paid, in that in such an approach children's rights are mainly associated with vulnerability and protection, and much less with agency and participation. Moreover, conceptual clarity on the relationship between human rights (including children's rights) and poverty is still missing (Doz Costa, 2008: 82). Finally, the limits and pitfalls of a rights-based approach should not be ignored (see e.g. Vandenhole,

---

[1] In the early 1990s, the reports of the European Observatory on National Policies to Combat Social Inclusion were written from this perspective; exclusion was defined in terms of the denial or non-realisation of social rights (Room, 1992). The General Report on Poverty ('Algemeen Verslag over de Armoede/Rapport Général sur la Pauvreté'), commissioned by the Belgian government and published in 1995 by the King Baudoin Foundation, also explicitly used rights language, as did Wresinski in his report *Grand pauvreté et précarité économique et sociale* in 1987.

2009a), and even explicitly addressed, in order to allow for an informed and strategic mobilisation of the children's rights framework.

In this chapter, the issue of children's rights and child poverty is primarily addressed from an analytical perspective, i.e. how children's rights (instruments) approach child poverty. Such an analytical approach is to be distinguished from an advocacy perspective, in which children's rights are invoked as an instrument of mobilisation and political pressure.

In what follows, the normative framework of children's rights as part of the broader human rights framework is succinctly presented. This part is intended to provide a very basic introduction for those not yet familiar with a rights perspective. In the second part, we examine the dominant understandings of the relationship between children's rights and poverty. The third part develops some tentative conclusions, in which attention is also paid to the policy relevance of some of the findings.

1. CHILDREN'S RIGHTS[2]

1.1. CHILDREN'S RIGHTS AS HUMAN RIGHTS

Children's rights are part of a broader body of human rights standards: children's rights are human rights of children. Children's rights should therefore not be studied nor invoked in isolation from the broader human rights framework.

Children's rights have been most comprehensively codified in the United Nations Convention on the Rights of the Child (CRC), which was adopted in 1989 after more than ten years of negotiations. Although children's rights cannot and should not be reduced to the CRC, the CRC is the most authoritative legally binding instrument on children's rights, thanks to its comprehensive scope and its nearly universal ratification by states (193 states have so far ratified the CRC; only Somalia and the United States have not done so yet). The focus in this chapter will therefore be on the CRC and its monitoring body, the Committee on the Rights of the Child.

The CRC is one of the currently nine core international human rights treaties; the others are the International Convention on the Elimination of All Forms of Racial Discrimination (1961), the International Covenant on Economic, Social and Cultural Rights (1966), the International Covenant on Civil and Political Rights (1966), the Convention on the Elimination of All Forms of Discrimination

---

[2] For a more extensive analysis, see e.g. Vandenhole, 2009b.

against Women (1979), the Convention against Torture and other Cruel, Inhuman and Degrading Treatment or Punishment (1984), the International Convention on the Protection of the Rights of all Migrant Workers and Members of their Families (1990), the Convention on the Rights of Persons with Disabilities (2006) and the Convention for the Protection of All Persons from Enforced Disappearance (2006). Except for the last, all have entered into force.

The CRC, being one of the core international human rights treaties, resembles the others on some points, while it also possesses its own unique characteristics. The CRC consists of three parts. Part I contains the definition of the child, general principles and obligations, and a detailed list of specific rights and obligations. Part II deals with the Committee on the Rights of the Child. Part III holds some final provisions on ratification, amendment, reservations and denunciation.

The CRC is comprehensive in scope, and covers civil, political, economic, social and cultural rights. Whereas the way they are listed emphasises their indivisibility, a reminder and casualty of the Cold War can be found in article 4 CRC, which differentiates the general obligation for the realisation of the rights, limiting it in the case of economic, social and cultural rights (ESC rights) to "the maximum extent of their available resources and, where needed, within the framework of international cooperation" (Cantwell, 1992: 27; Alston, 1994: 7). This relative obligation with regard to the ESC rights of children reminds us of the long-lasting discussion on the legal nature of ESC rights. More often than not, they are considered second-class rights in jurisprudence and politics. The acceptance that the realisation of economic, social and cultural rights can be achieved only progressively and subject to the availability of resources bears out the recognition of a time dimension and a need for priorities in light of resource constraints. However, as the Committee on Economic, Social and Cultural Rights (CESCR) has clarified, this recognition does not grant states a *carte blanche*: they must begin immediately to take steps in order to realise ESC rights as expeditiously as possible. States also need to adopt a strategy and plan of action to realise these rights. Prioritisation too is subjected to process and substance conditions. The process in which prioritisation is decided needs to be participatory and transparent: all stakeholders, including the poor, must be involved so that all segments of society, especially the poor, can express their value judgements regarding priorities. Substantively, core obligations (i.e. obligations pertaining to the minimum essential levels of the rights) should always be complied with. Moreover, there is a strong presumption that retrogressive measures are not permissible (OHCHR, 2004: 22–27; CESCR 1990 and 2002).

A popular categorisation among children's rights proponents and CRC commentators is the three Ps: rights to protection, provision and participation. The downside of this categorisation is not only that it departs from the one human rights actors are familiar with, but also and more importantly that the term "provision rights" tends to reinvigorate the outdated misunderstanding or misrepresentation that economic and social rights are exclusively about provision. Meanwhile, it has been widely accepted that obligations relating to ESC rights can be best understood as obligations to respect, to protect and to fulfil, and that the latter obligation consists of sub-obligations to facilitate, to promote and to provide. Only the sub-obligation to fulfil-provide requires considerable mobilisation of resources. The obligation to respect, on the other hand, is of immediate effect. It only requires abstention from certain conduct and does not necessitate any resources (Vandenhole, 2003: 443–444).

The CRC is governed by four overarching general principles, which are explicitly reflected in four provisions: the right to equality and non-discrimination (art. 2 CRC); the best interests of the child (art. 3 CRC); the right to life, survival and development (art. 6 CRC); and respect for the views of the child, sometimes also referred to as the right to participation (art. 12 CRC). These general principles pervade the whole Convention and are to be taken into account when interpreting a provision in the CRC.

The prohibition of discrimination in the field of economic, social and cultural rights can be argued not to be limited by article 4 CRC: it is of immediate effect. The "best interests of the child" is a relative new interpretation principle in international law, which was introduced to it by the CRC (Freeman, 2007: 1). Its meaning and application is very problematic. The interests of the child can be understood in different ways: as basic interest, developmental interest or autonomy interest (Eekelaar, 1992: 230–231), or also in light of children's needs, potential harm to children or their wishes and feelings (Freeman, 2007: 31). In light of the interdependence and interaction between different rights in the CRC, it is submitted that the best interests of the child should be understood in light of all other rights and general principles, so that children themselves should have a say in defining what is in their interest. It is noteworthy that the best interests of the child are only "*a* primary consideration", not *the* primary or paramount consideration. The best interests of the child are therefore to be balanced with other interests (Freeman, 2007: 60).

The right to life, survival and development is unique in its formulation. Other core human rights treaties protect the right to life only, without mentioning survival and development. The reference to survival was intended to emphasise the positive obligations incumbent on states parties to prolong children's lives. Survival is closely related to the healthy development of children and thereby introduces obligations of fulfilment (Nowak, 2005: 12–14 and 36–37). The attention paid to the development of children, to be understood holistically (CRC Committee, 2003: para. 12), is closely related to the concept of human development as advocated by the World Health Organization and UNICEF in the 1980s (Nowak, 2005: 7 and 14). Finally, the right to participation is a cluster of rights, with at its core the right to express one's view and the right to have that view taken into account. The right to express one's views is limited to children who are capable of forming their views and extends only to matters that affect them. The right to have the views expressed taken into account is qualified by references to age and maturity. Quite often, the right to participation is balanced against the best interests of the child (Ang and others, 2006: 14 and 18).

1.2. IMAGES OF CHILDREN AND CHILDHOOD IN THE CRC

Two schools of thought or perspectives on children have informed the CRC. On the one hand, there is the view that children need special protection and priority care. That was the almost exclusive theme of the 1924 and 1959 Declarations, which should be understood in light of the two World Wars (Cantwell, 1992: 19). This view has been referred to as the biomedical model of childhood: "children as passive victims who are psychologically scarred and vulnerable" (Hinton, 2008: 288). On the other hand, there are proponents of recognizing children as autonomous individuals and "fully-fledged beneficiaries of human rights" (Cantwell, 1992: 27). In non-legal terms reference is made to "children's agency, resilience and coping mechanisms" (Hinton, 2008: 288). It may be argued that the following balance has been struck between these schools: "The CRC sees the child as an initially highly vulnerable person in need of protection, nurturing and care who under parental guidance gradually prepares for an independent life in a social setting of rights and duties when reaching eighteen" (Eide, 2006: 3).

The CRC Committee has grouped a substantive number of articles under the heading of special protection measures. Some pertain to *specific groups* of children, such as children seeking refugee status (art. 22) and children belonging to minorities or of indigenous origin (art. 30). Other provisions hold guarantees for *all* children against certain risks such as economic exploitation, harmful work, illicit use of narcotic drugs and psychotropic substances, sexual exploitation and abuse, abduction, sale and trade, or any other form of exploitation (arts. 32–36). States also undertake to respect and to ensure respect for the rules of international humanitarian law in armed conflict (art. 38). Article 39 imposes the obligation on states to promote the physical and psychological recovery and social reintegration of child victims.

2. CHILDREN'S RIGHTS AND POVERTY

2.1. HUMAN RIGHTS AND POVERTY – IS POVERTY A VIOLATION OF HUMAN RIGHTS?

There is the assumption in the human rights community that poverty can or should be one of the underlying and recurrent concerns. The relationship between human rights (including children's rights) and poverty is conceptually vague, though.

Poverty is neither explicitly mentioned in any of the other core international human rights treaties, nor mentioned as such in the CRC. Moreover, a rights approach is characterised by certain conceptual constraints. First, when addressing poverty from the perspective of human rights law, a statist perspective is taken. Human rights treaties, including the CRC, focus mainly on *state* obligations. Technically speaking, poverty therefore only constitutes a *violation* of human rights, including children's rights, if the state has failed to abide by its obligations of respect, protection or fulfilment. Human rights law also tends to take an individualistic approach. The lack of a clear reference right for addressing poverty, in combination with the statist and individualistic bias of human rights law, may lead therefore to a fragmented approach in which certain issues are singled out (social assistance or housing or medical assistance), but not comprehensively taken up. Moreover, root causes and structural causes may remain unaddressed. Riedel, who is one of the prominent members of the Committee on Economic, Social and Cultural Rights, has submitted that "the Committee deals with the issue of poverty through looking at its adverse effects on the realisation of *specific* rights covered by the ICESCR and the vulnerability of specific groups" (my emphasis) (Report Open-Ended Working Group, 2004: para. 45).

A further challenge is posed by the principle of progressive realisation of economic, social and cultural rights. The realisation of ESC rights is understood to be qualified by resource availability (see above). These rights are in principle not to be realised immediately. They can be realised progressively, subject to the maximum use of all available resources. Sensitive questions often arise as to who has the legitimacy to assess whether maximum use has been made of available resources, the executive or the judiciary. In any case, a state's failure to address poverty and social exclusion does not automatically amount to a violation (Vandenhole, 2008).

Several approaches can link children's rights/human rights to poverty.[3] In what follows, we outline two: a right to protection against poverty and a rights-based

---

[3] For an extensive discussion and typology, see Doz Santos, 2008.

approach to poverty. We then turn to the Committee on the Rights of the Child's approach to poverty.

2.1.1. A right to protection against poverty

Only a few human rights instruments mention poverty explicitly. We will deal with two European ones here. In the Revised European Social Charter (RESC, 1996), a Council of Europe instrument, the right to protection against poverty and social exclusion is guaranteed. Article 30 RESC reads:

With a view to ensuring the effective exercise of the right to protection against poverty and social exclusion, the Parties undertake:

a   to take measures within the framework of an overall and co-ordinated approach to promote the effective access of persons who live or risk living in a situation of social exclusion or poverty, as well as their families, to, in particular, employment, housing, training, education, culture and social and medical assistance;
b   to review these measures with a view to their adaptation if necessary.

In para. 114 of the Explanatory Report, poverty and social exclusion are explained in broad terms, as follows:

The term "poverty" in this context covers persons who find themselves in various situations ranging from severe poverty, which may have been perpetuated for several generations, to temporary situations entailing a risk of poverty. The term "social exclusion" refers to persons who find themselves in a position of extreme poverty through an accumulation of disadvantages, who suffer from degrading situations or events or from exclusion, whose rights to benefit may have expired a long time ago or for reasons of concurring circumstances. Social exclusion also strikes or risks to strike persons who without being poor are denied access to certain rights or services as a result of long periods of illness, the breakdown of their families, violence, release from prison or marginal behaviour as a result for example of alcoholism or drug addiction.

The RESC, in the interpretation given by the European Committee of Social Rights (ECSR), clearly breaks away from the dispensing of charity to the recognition of rights, which is also evidenced by the approach taken in article 13 on the right to social and medical assistance.

As to the substance of the right to protection against poverty and social exclusion, no guarantee of minimum resources is mentioned, as that is covered by article 13, which provides for a right to social assistance. Instead, states commit themselves to a comprehensive and coordinated approach to the alleviation of poverty and social exclusion. Measures included may or may not imply financial benefits. The Explanatory Report emphasises that "States subscribing to this provision are encouraged to restrict financial benefits to those who cannot help themselves by their own means" (para. 116).

The ECSR, which monitors the RESC, has established a link between poverty and social exclusion on the one hand and human dignity on the other hand by qualifying poverty and social exclusion as a violation of the dignity of human beings (though not as a violation of human rights strictly speaking). Nevertheless, it limits poverty to deprivation due to a lack of resources, although it does recognize "the multidimensional phenomena of poverty and social exclusion". In the Committee's view, the obligation to adopt an overall and co-ordinated approach requires the elaboration of an analytical framework, a set of priorities, measures to prevent and remove obstacles to accessing fundamental social rights, and inclusive or participative monitoring mechanisms. An increase in the resources deployed is required as long as poverty and social exclusion persist. Where necessary, measures should specifically target the most vulnerable groups, one of which children are often considered to be. In sum, analytically, poverty is understood as mainly income-related and is considered to be a violation of human dignity. The response required of states consists of strengthening access to social rights, including by allocating adequate resources to the approach taken. Targeted measures are expected for the most vulnerable groups, such as children.

The Charter of Fundamental Rights of the European Union too refers explicitly to social exclusion and poverty. Article 34(3) on social security and social assistance, which is part of the chapter on solidarity, reads:

> *In order to combat social exclusion and poverty, the Union recognises and respects the right to social and housing assistance so as to ensure a decent existence for all those who lack sufficient resources, in accordance with the rules laid down by Community law and national laws and practices.*

Rather than recognizing a right to protection from poverty, the Charter acknowledges that in policies combating social exclusion and poverty a right to social and housing assistance is crucial to ensuring a decent existence for poor people. The Convention that drafted the Charter has explained that this provision draws on articles 30 and 31 RESC. The Convention also submitted that the provision should particularly be taken into account when adopting measures aimed at encouraging cooperation among member states in combating social exclusion (art. 137 (2), last sub-paragraph EC Treaty). The Charter was adopted as a legally non-binding instrument in 2000, but readopted after some adaptations in 2007. In accordance with article 1, 8) of the 2007 Treaty of Lisbon amending the EC and the EU Treaties, the 2007 Charter has become binding on the EU since the entry into force of the Treaty of Lisbon on 1 December 2009.

2.1.2. Human rights-based approaches to poverty

An analysis through the prism of human rights or children's rights is often coined "a (human) rights-based approach ((H)RBA)". There is no single approach or understanding, but rather a multiplicity of (human) rights-based approaches to poverty and/or development. Common to all (H)RBAs, nonetheless, is their focus not only on the end result or outcome (which should be the realisation of human rights), but also on the overall process. Central features of most (H)RBAs can be summarised in the acronym PANEL, i.e. participation, accountability, non-discrimination, empowerment and linkage to human rights. (H)RBAs typically approach poverty from a global perspective, so as both a domestic and a global issue (i.e. beyond states' own borders). Very often, global north-south relations and the issue of development cooperation are in the background.

CESCR, which monitors the International Covenant on Economic, Social and Cultural Rights, adopted a (non-binding) Statement on poverty and the Covenant in 2001. In that Statement, the Committee seeks to clarify "the distinctive contribution of international human rights to poverty eradication", or in other words, "how human rights generally, and the Covenant in particular, can empower the poor and enhance anti-poverty strategies" (CESCR, 2001: para. 3).

The Committee argues that an understanding of poverty's broader features beyond income poverty (a "multidimensional understanding of poverty") corresponds with "numerous provisions of the Covenant" (para. 7). The Committee homes in on all key features of a rights-based approach, i.e. participation, accountability, non-discrimination, empowerment and linkage to human rights, which it considers to be "essential elements of anti-poverty strategies". Drawing on the World Bank's *Voices of the Poor*, the Committee's starting point is that the "common theme underlying poor people's experiences is one of powerlessness". Given the "empowering potential of human rights", the "challenge is to connect the powerless with the empowering potential of human rights" (para. 6). Human rights are believed to be able to contribute to equalising the distribution and exercise of power, as they impose legal obligations on duty-holders. While the Committee attaches central importance to the right to an adequate standard of living (art. 11), it emphasises that other economic, social and cultural rights, as well as civil and political rights, are "indispensable to those living in poverty" (paras. 8–10). In other words, all human rights are equally important.

Non-discrimination and equality norms are understood to require particular attention for vulnerable groups and individuals from such groups (para. 11). Children growing up in poverty are believed to be often permanently disadvantaged (para. 5). The importance of the right to participation of those who are affected (i.e. those who live in poverty) is mainly seen in light of its effectiveness. Substantively, while free and fair elections are believed to be a crucial component of the right, they do not suffice. Active and informed participation in anti-poverty policies or programmes is needed (para. 12). Finally, the Committee calls for accessible, transparent and effective accountability mechanisms in order to be able to hold all duty-bearers to account (para. 14).

In sum, in CESCR's view international human rights are believed to add important value to anti-poverty policies, which are said to be "more likely to be effective, sustainable, inclusive, equitable and meaningful to those in poverty if they are based upon international human rights" (para. 13). At the same time the Committee accepts that "human rights are not a panacea", and accords rather a facilitative role to human rights ("they can *help* to equalise the distribution and exercise of power within [...] societies" (emphasis added) (para. 6). The Committee departs from the technical understanding that human rights obligations are only incumbent on states: it refers to duty-holders more generally, among them states and international organisations. No clarification is offered, however, as to who the other duty-holders are. Moreover, no legal accountability mechanisms seem to be required for duty-holders other than the state (para. 14).

In 2004 and 2006, the Office of the High Commissioner for Human Rights (OHCHR) issued two publications on a human rights-based approach to Poverty Reduction Strategies (PRS). These publications drew on reports prepared by three experts, Nowak and Hunt (both with a legal background in human rights) and Osmani (who is a development economist). The 2004 publication is a study of the conceptual framework for the integration of human rights into poverty reduction strategies. The 2006 publication contains principles and guidelines for a human rights approach to poverty reduction strategies.

In the conceptual framework publication of 2004, the concept of poverty is not surprisingly explicitly addressed, albeit with a certain purpose in mind. The aim is to find a conceptual bridge between the discourses on poverty and human rights. A definition of poverty is therefore needed that refers to non-fulfilment of human rights without delinking it from the well-established connection with deprivation caused by economic constraints (OHCHR, 2004: 5–6). That conceptual bridge is found in Sen's capability approach. Sen defines capability as a person's freedom or opportunities to achieve wellbeing. A poor person has very restricted opportunities to pursue his or her wellbeing (i.e. a low level or failure of basic capabilities), which has at least partly to do with an inadequate command over economic resources (income, publicly provided goods and services, etc) (OHCHR, 2004: 6–9). In the report, poverty as the absence or inadequate realisation of certain basic freedoms (in the words of the capability approach) is believed to correspond to the non-fulfilment of rights to basic freedoms (in human rights terminology) (OHCHR, 2004: 10).

A major criticism of this assumption of conceptual equivalence between basic capabilities and human rights has been that the notion of capabilities is essentially contingent – i.e. what is basic may vary considerably from one society to another, while human rights are essentially universal. Doz Santos therefore fears that the proposed conceptual equivalence may compromise conceptual developments in human rights law towards the clarification of state obligations in the field of ESC rights (Doz Santos, 2008: 87–88). While this fear may be understandable from a technical legal perspective, it seems to focus too strongly on what distinguishes both frameworks rather than on what unites them.

The nexus between human rights and poverty is explained in the 2004 report in terms of constitutive, instrumental and constraint-based relevance. Constitutive relevance means that non-fulfilment of those human rights that correspond to basic capabilities in a given society, and in which inadequate command over economic resources plays a causal role, counts as poverty. The instrumental relevance of human rights for poverty lies in the ability of human rights to promote the cause of poverty reduction. Finally, its constraint-based relevance implies that human rights constrain the types of action that are permissible in poverty reduction policies and actions (OHCHR, 2004: 10–12).

The 2004 report explains the added value of a human rights approach to poverty reduction strategies by invoking its key features, as summarised by the acronym PANEL. The conclusion drawn is that human rights help to ensure that the key concerns of poor people also are the key concerns of poverty reduction strategies. There is, however, some sense of realism in that it is admitted that empowerment through human rights can only take place in practice to the extent that poor people are able to access and enjoy human rights. The caveat of accessibility is also mentioned with regard to accountability mechanisms. Moreover, when it comes to participation it is recognized that the participation of poor people is "deeply dependent" on other human rights. So while conceptually the added value of a (human) rights-based approach is not questioned, the challenges of putting it into practice or not are ignored (OHCHR, 2004: 13–20). No specific attention is paid to child poverty.

The 2006 OHCHR publication on guiding principles and guidelines, which builds on the conceptual paper, does not explore the concept of poverty or the nexus between human rights and poverty. It does pay some attention to children. Children are explicitly mentioned as victims (OHCHR, 2006: paras. 109 en 127, with regard to child labour) and as belonging to groups in a particularly vulnerable situation (paras. 135 and 203). The prioritisation and provision of tailor-made services for children are recommended in the areas of health care and education (paras. 179 and 191). And the importance of education for escaping from poverty is acknowledged (para. 184).

Draft guiding principles "extreme poverty and human rights: the rights of the poor" (DGP) were adopted by the (since defunct) Sub-Commission on the Promotion and Protection of Human Rights in August 2006. After two rounds of consultation, a seminar was held on the DGPs in January 2009. The DGPs simply align themselves with the 2001 Statement of CESCR on the concept of poverty. Nor is there much conceptual analysis of the relationship between human rights and poverty beyond the submission that extreme poverty and exclusion constitute a violation of human dignity and a negation of human rights (10th preambular para. and para. 2 DGP). The ad hoc working group that prepared the DGP focused on the right to life and qualified poverty in general as a violation of the enjoyment of ESC rights, and extreme poverty as a gross violation of the right to life and human rights (Bengoa, 2002). Central principles in the DGP are participation by the poor in decision-making and implementation and the state's obligation to empower people living in poverty to organise themselves.

Finally, Sepúlveda, in 2008 appointed UN independent expert on the question of human rights and extreme poverty, outlined the conceptual framework she will work with in her first report. She aligns herself with a multidimensional working definition of extreme poverty, and characterises the relationship between human rights and poverty very much in line with the recently established position of the UN human rights community. She pays explicit attention to children. Analytically, she believes that children are differently affected by poverty. Normatively, she emphasises the need to ensure the best interests of the child and participation in policy and implementation measures (Sepúlveda, 2008: paras. 36–38).

2.2. THE CRC AND THE COMMITTEE ON THE RIGHTS OF THE CHILD AND POVERTY

The CRC does not contain a specific reference to poverty, let alone a right to protection against poverty. Nor has the Committee on the Rights of the Child so far adopted a general comment in which it extensively addresses poverty.

The Committee, in line with the understanding of CESCR, considers that among children there are groups of children with heightened vulnerabilities ("most vulnerable groups of children"), among them poor children (CRC Committee, 2005: para. 24). There is both recognition of the particular vulnerability *to* poverty and deprivation of some groups of children – such as young children

(CRC Committee, 2005: paras. 2 (f) and 36), indigenous children (CRC Committee, 2009: paras. 34 and 70), immigrant children (see, e.g., CRC Committee, Concluding Observations France, para. 78; Concluding Observations Sweden, para. 52) and children living in single-parent households (CRC Committee, Concluding Observations Sweden, para. 52) – and of the particular risks to young children that result *from* poverty and social exclusion (CRC Committee, 2005: para 8). The conclusion drawn from the particular vulnerability of young children to and resulting from poverty is that proper prevention and intervention strategies during early childhood may impact positively on young children's current well-being and future prospects. In the allocation of the maximum available resources, there needs to be a special focus on eradicating poverty and reducing inequalities (CRC Committee, Concluding Observations France, para. 19).

Rights in the CRC that may be mobilised in a children's rights approach to poverty may be article 2 (non-discrimination) and articles 26 and 27 in particular. Discrimination and exclusion seem to be considered as being at the origin of poverty (CRC Committee, Concluding Observations France, para. 78). Article 26 guarantees the right to benefit from social security, while article 27 provides for a right to a standard of living that is adequate for the child's mental, spiritual, moral and social development. In Eide's understanding, this right to an adequate standard of living goes "beyond the purely material aspects of living such as food and housing", as the "standards, or the conditions under which the child lives, must be adequate for the child's physical, mental, spiritual, moral and social development" (Eide, 2006: 17).

The division of responsibility outlined in article 27 goes beyond a traditional understanding of human rights in a vertical relationship between the individual (in this case the child) and the state. For primary responsibility to secure the conditions of living necessary for the child's development lies with the parents, albeit within their abilities and financial capacities (see art. 27, para. 2). The corresponding obligation of the state is to ensure that parents fulfil their obligations towards their child (obligation to protect). This attribution of primary responsibility to the parents builds on the Convention's "conception of the ideal type of setting for the upbringing of the child: a family with the will and the capability to care for the child during its many years, starting with the pregnancy and from birth to full maturity at the age of eighteen" (Eide, 2006: 3). It goes without saying that parents' capacity to live up to this ideal-type conception is strongly dependent on their resources, material and non-material. If, despite their efforts, parents are unable to ensure proper conditions for the development of the child, the state has an obligation to assist (Eide, 2006: 3 and 26).

This subsidiary obligation placed on states in para. 3, to assist parents in the implementation of the right, is limited in several ways, however. First of all, it is subject to the means available to the state. Secondly, the state is to intervene only in case of need, and its intervention seems to remain limited to material needs: the state is to provide material assistance and support programmes, particularly with regard to nutrition, clothing and housing (obligation to fulfil) (Eide, 2006: 2). In its concluding observations, the Committee seems to go beyond this material assistance and urges states parties "to render appropriate assistance to parents and legal guardians in the performance of their child-rearing responsibilities, in particular for families in crisis situations due to poverty, absence of adequate housing or separation" (CRC Committee, Concluding Observations France, para. 60). Nevertheless, it addresses child poverty most explicitly under the heading of "standard of living" (CRC Committee, Concluding Observations France, paras. 78 and following).

2.3. ARE CHILDREN'S RIGHTS BIASED AGAINST THE POOR?

The CRC Committee has recognised the negative impact on young children of growing up in relative or absolute poverty (CRC Committee, 2005: para. 26). Which consequences should be inferred from this recognition? People living in poverty often experience children's rights as being played out against them, in particular in order to justify placement into care. Euronet confirms this experience: "children from families suffering extreme poverty and deprivation are placed in institutional care, instead of supporting families at risk through better services and benefits". The CRC Committee came to the same conclusion in its recent concluding observations on France, in which it noted that children were placed in alternative care as a result of low parental income and that new draft legislation on the national adoption of children in a situation of parental neglect involved a definite risk of separating children, especially those from low income families and families living in poverty, from their family environment (CRC Committee, Concluding Observations France, paras. 62 and 65). While it is true that children's rights are mobilised against parents living in poverty, particularly to justify placement into care, neither the CRC nor the Committee favour such an instrumentalisation. In cases of neglect, the state has an obligation to intervene in order to offer protection and care (art. 3 §2 and art. 19 CRC).

However, one of the reasons for allocating primary responsibility to the parents in art. 27 on the right to an adequate standard of living (and for upbringing and development more generally, compare also art. 18 CRC) was precisely to protect parents against excessive state intervention (Detrick, 1992: 459). Moreover, separation from parents is only permissible in the child's best interests (art. 9

CRC), as it is accepted that the family usually provides the best setting for a child's development (art. 5 CRC) (compare CRC Committee, 2005: para. 15). Similarly, it is now established case-law of the European Court of Human Rights that in light of the right (of parents) to family life, children can only be removed from their family as a measure of last resort, and for the shortest possible period. Family reunion should be the ultimate aim of any placement into care. Nevertheless, in the child's best interests family reunion may sometimes not occur. A central concept is therefore the best interests of the child. This remains, however, a vague and versatile concept, which tends to be mobilised to justify whatever decision is taken with regard to a child.

3. SOME TENTATIVE CONCLUSIONS WITH POTENTIAL POLICY RELEVANCE

Overall, poverty in Western countries in the North has not yet been analysed very thoroughly or systematically from a human rights/children's rights perspective. Only tentative conclusions can be made.

The multidimensional nature of poverty is recognised in human rights-based approaches to poverty. Nevertheless, income poverty remains central in recommendations for remedial action, as illustrated by the emphasis on strengthening access to social rights, including by allocating adequate resources. Poverty is considered to create a specific vulnerability amongst children. A positive consequence of this is that because of that heightened vulnerability, they are believed to be in need of special attention and care. A danger, however, particularly in practice, is that this need for protection against neglect tends to be invoked for disproportionately more frequent placement into care of children living in poverty.

A rights-based approach to poverty is believed to have empowering potential, in particular because of the shift from charity to rights, with the concomitant principles of participation, accountability, non-discrimination, empowerment and linkage with human rights. Clearly, by invoking a legally binding normative framework, ideologically loaded discussions can be evaded. Inserting the (H)RBA principles into policies and programmes also narrows policy discretion, in that a minimum threshold is established; process requirements such as participation and prioritisation of the poor, and of poor children in particular, are introduced; and the best interests of the child are taken into account.

It is not clear, however, whether and to what extent the underlying assumption in human and children's rights law of the autonomy of the individual involves a risk of individuating and decontextualising poverty. Moreover, the conceptual advantages of a (H)RBA do not materialise automatically into practice. First of all, in a (H)RBA the state's responsibility with regard to poverty is a subsidiary one, which is moreover centred on material conditions and access to social services. One may wonder whether this approach corresponds to current understandings of the nature of poverty.

Secondly, lessons may be learnt from the debate in development studies, to which issues of power and participation have been central (Hinton, 2008: 289). Drawing on the critique of participatory development (for adults), the inherent limitations of child participation if the existence of (unequal) power relations and their impact are ignored, have been pointed out (Hart, 2008: 407). So, further clarification is needed of the empowering potential of the human and children's rights approach. In particular, the question can be asked whether human and children's rights do not presuppose rather than "create" empowered, participating individuals who are able and willing to invoke their rights (if necessary before a court of law) in order to hold the state accountable. Caution may also be warranted so as not to place too much responsibility on those who may not have the power to change their circumstances (Hinton, 2008: 287, summarising Moncrieffe, 2004). However, commonly held assumptions about power, in which the emphasis tends to be placed on those who wield it, and about those who are powerless (in our case: the poor, and in particular poor children) have also been challenged. Gallagher submits that (governmental) power is ambivalent, for it is both a means of control and of resistance. Power is on the one hand used to persuade "people to participate in their own subjection", but it also "equips [a human being] to become an independent actor". In his view, "participation has the potential for both compliance and insubordination" (Gallagher, 2008: 401–402). One may wonder whether and to what extent poverty impinges on children's ability not only to resist power, but also to change power relations and the concomitant social exclusion.

REFERENCES

ALSTON, Ph. (1994), The Best Interests Principle: Towards a Reconciliation of Culture and Human Rights, *International Journal of Law and the Family*, (8): 1–25.

ANG, F., BERGHMANS, E., CATTRIJSSE L. et al. (2006), Participation Rights in the UN Convention on the Rights of the Child, in: F. ANG, E. BERGHMANS, L. CATTRIJSSE et al., *Participation Rights of Children*. Antwerp: Intersentia, 2006, 9–26.

BENGOA, J. (2002), *Poverty and human rights. Programme of work of the ad hoc working group established to prepare a study to contribute to the drafting of an international declaration on extreme poverty and human rights*, UN Doc. E/CN.4/Sub.2/2002/15 of 25 June 2002.

CANTWELL, P. (1992), The Origins, Development and Significance of the United Nations Convention on the Rights of the Child, in: S. DETRICK, *The United Nations Convention on the Rights of the Child: A Guide to the "Travaux Préparatoires"*. Dordrecht/Boston/London: Martinus Nijhoff, 19–30.

CESCR (1990), *General Comment No. 3: The nature of States parties obligations (Art. 2, par. 1)*, UN Doc. E/1991/23 of 14 December 1990.

CESCR (2001), *Statement on Poverty and the International Covenant on Economic, Social and Cultural Rights*, UN Doc. E/C.12/2001/10 of 10 May 2001.

CESCR (2002), *General Comment No. 15, The right to water (arts. 11 and 12)*, UN Doc. E/C.12/2002/11 of 20 January 2003.

CRC COMMITTEE (2003), *General Comment No. 5, General measures of implementation*, UN Doc. CRC/GC/2003/5 of 3 October 2003.

CRC COMMITTEE (2005), *General Comment No. 7, Implementing child rights in early childhood*, UN Doc. CRC/C/GC/7/Rev.1 of 20 September 2006.

CRC COMMITTEE (2009), *General Comment No. 11, Indigenous children and their rights under the Convention*, UN Doc. CRC/C/GC/11 of 12 February 2009.

CRC COMMITTEE, Concluding Observations France, UN Doc. CRC/C/FRA/CO/4 of 11 June 2009.

CRC COMMITTEE, Concluding Observations Sweden, UN Doc. CRC/C/SWE/CO/4 of 12 June 2009.

DE ALBUQUERQUE, C., *Report of the open-ended working group to consider options regarding the elaboration of an optional protocol to the International Covenant on Economic, Social and Cultural Rights on its first session (Geneva, 23 February-5 March 2004)*, UN Doc. E/CN.4/2004/44 of 14 March 2004.

DETRICK, S. (1992), *The United Nations Convention on the Rights of the Child: A Guide to the "Travaux Préparatoires"*. Dordrecht/Boston/London: Martinus Nijhoff.

DOZ COSTA, F. (2008), Poverty and Human Rights: from Rhetoric to Legal Obligations. A Critical Account of Conceptual Frameworks, *SUR – International Journal on Human Rights*, (5): 9, 81–106.

EEKELAAR, J. (1992), The Importance of Thinking That Children Have Rights, *International Journal of Law and the Family*, (6): 1, 221–235.

EIDE, A. (2006), Article 27: the Right to an Adequate Standard of Living, in: A. ALEN, J. VANDE LANOTTE, E. VERHELLEN, F. ANG, E. BERGHMANS & M. VERHEYDE (eds.), *A Commentary on the United Nations Convention on the Rights of the Child*. Leiden/Boston: Martinus Nijhoff.

FREEMAN, M., Article 3: the Best Interests of the Child, in: A. ALEN, J. VANDE LANOTTE, E. VERHELLEN, F. ANG, E. BERGHMANS & M. VERHEYDE (eds.), *A Commentary on the United Nations Convention on the Rights of the Child*. Leiden/Boston: Martinus Nijhoff, 2007.

GALLAGHER, M. (2008), Foucault, Power and Participation, *International Journal of Children's Rights*, (16): 3, 395–406.

HART, J. (2008), Children's Participation and International Development: Attending to the Political, *International Journal of Children's Rights*, (16): 3, 407–418.

HINTON, R. (2008), Children's Participation and Good Governance: Limitations of the Theoretical Literature, *International Journal of Children's Rights*, (16): 3, 285–300.

KING BAUDOIN FOUNDATION (1995), *Algemeen verslag over de armoede. Armen over hun situatie in België en wat eraan kan worden gedaan. Een opdracht op vraag van de overheid*, available at www.kbs-frb.be/publication.aspx?id=177842&LangType=2067.

MONCRIEFFE, J. (2004), *Power Relations, Inequality and Poverty*. Brighton: IDS.

NOWAK, M. (2005), Article 6: the Right to Life, Survival and Development, in: A. ALEN, J. VANDE LANOTTE, E. VERHELLEN et al. (eds.), *A Commentary on the United Nations Convention on the Rights of the Child*. Leiden/Boston: Martinus Nijhoff.

OHCHR (2004), *Human Rights and Poverty Reduction – A Conceptual Framework*, Geneva: United Nations.

OHCHR (2006), *Principles and Guidelines for a Human Rights Approach to Poverty Reduction Strategies*, Geneva: United Nations.

ROOM, G. et al. (1992) *Observatory on National Policies to Combat Social Exclusion. Second Annual Report*, Brussels, Directorate general For Employment, Social Affairs and Industrial relations, Commission of the European Communities.

SEPÚLVEDA CARMONA, M. (2008), *Report of the independent expert on the question of human rights and extreme poverty*, UN doc. A/63/274 of 13 August 2008.

VANDENHOLE, W. (2003), Completing the UN Complaint Mechanisms for Human Rights Violations Step by Step: Towards a Complaints Procedure to the International Covenant on Economic, Social and Cultural Rights, *Netherlands Quarterly of Human Rights*, (21), 423–462.

VANDENHOLE, W. (2008), Conflicting Economic and Social Rights: The Proportionality Plus Test, in: E. BREMS (ed.), *Conflicts Between Fundamental Rights*. Antwerp: Intersentia, 559–589.

VANDENHOLE, W. (2009a), The Limits of Human Rights Law in Human Development, in: E. CLAES, W. DEVROE & B. KEIRSBILCK (eds.), *Facing the Limits of the Law*. Berlin/Heidelberg: Springer, 355–374.

VANDENHOLE, W. (2009b), The Convention on the Rights of the Child, in: K. DE FEYTER & F. GOMEZ ISA (eds.), *International Human Rights Law in a Global Context*. Bilbao: University of Deusto, 451–472.

WRESINSKI, J. (1987), *Grande pauvreté et précarité économique et sociale*, available at www.atd-quartmonde.org/IMG/pdf/WRES_JO87.pdf

3. CHILD POVERTY, SOCIAL EXCLUSION AND CHILDREN'S RIGHTS: A VIEW FROM THE SOCIOLOGY OF CHILDHOOD

Virginia MORROW

INTRODUCTION

Discussions and debates about child poverty and children's rights rarely seem to connect, and pass each other like trains on parallel tracks. While well-established arguments are made that poverty and social exclusion breach human rights (Room, 1999; Dean, 2007), the links between *children's* rights as enshrined in the UN Convention on the Rights of the Child (1989, ratified by the UK government in 1991; henceforth UN CRC) and *child* poverty are not so clearly articulated in research. This is particularly the case in the UK, where it is not unusual to find studies of childhood poverty based on children's views and accounts, conducted in the 1990s (Middleton et al., 1997) and early 2000s (Ridge, 2002), without seeing a single reference to the UN CRC. When child poverty specialists do attempt to include a discussion of the UN CRC in their work, they tend to do so in a partial way, selecting various articles of the Convention to fit their argument, and there is a danger of associating children in poverty with 'victimhood' (Hill et al., 2006; see also Morrow & Mayall, 2009, 2010, for a critique of UNICEF 2007).

Why is there this gap? This chapter attempts to explore some of the reasons through the lens of the 'new' sociology of childhood, based on a brief review of available literature and drawing upon some empirical research by the author, and suggests some ways forward. The chapter is structured as follows. The first section introduces the sociology of childhood. The second section makes some general points about child poverty research in the UK. Section 3 outlines the UN CRC and how it intersects with child poverty. Section 4 describes social exclusion and human capabilities approaches to child poverty. The fifth section draws on an example of an empirical study conducted by the author, framed within the sociology of childhood, to try to demonstrate the linkages between children's experiences of social exclusion and to explore implications for their rights. The final section discusses children's participation and the implications for poverty debates.

The chapter does not explore the contribution to debates about child poverty or children's rights from mainstream (as opposed to radical or critical) developmental psychology and child psychiatry. Indeed, in the UK at any rate, developmental psychology is remarkably silent/apolitical on questions of children's rights, focusing on 'needs' instead of entitlements. Nor does the chapter focus on the very large literature on inequality, social stratification and social class that exists in the UK – again, this body of work tends not to be based on, or engage with, debates about children's rights.

1. THE SOCIOLOGY OF CHILDHOOD

The sociology of childhood grew out of dissatisfaction with the general neglect of children within sociology, both at the macro level and at the micro level. The Danish sociologist Jens Qvortrup was the first to break with the sociology of the family in the mid-1980s (Qvortrup, 1985), pointing out that there were plenty of 'sociologically relevant discussions of children's problems and problem children' (Qvortrup, 1987: 3) but a dearth of studies that considered childhood as a structural segment of populations. Few studies were grounded in children's experiences of their daily lives, and during the late 1980s and early 1990s it was increasingly acknowledged that *sociological* research (as opposed to psychological/behavioural/medical research) with children was underdeveloped (Qvortrup, 1987; Prout & James, 1990). The new paradigm involved moving on from the narrow focus of socialisation and child development (the study of what children will *become*) to a sociology that attempts to take children seriously as they experience their lives *in the here-and-now* as children.

Alanen (2001; see also Mayall, 2002) has identified three internally related 'sociologies of childhood'. The first is sociologies of children, in which children are understood as agents and as participants in constructing knowledge. This involves:

> *Studying children in their own right and from their own perspectives, ... taking children as the units of research and focusing the study directly on children and their life conditions, activities, relationships, knowledge and experiences (Alanen, 2001: 12).*

These construct an account of childhood and emphasise 'the present tense of childhood' (Mayall, 2002) (instead of future becomings). The second is what Alanen refers to as 'the deconstructive sociology of childhood', emphasising how ideas about childhood change through time and space (see also Prout & James, 1990; Hendrick, 1990, 2003) and how children's roles and activities differ according to culture, history, class, dis/ability, gender, ethnicity and so on. 'The task of the sociologist is to deconstruct ... cultural ideas, images, models and practices of children and childhood' (Alanen, 2001: 13). The analytic function of deconstruction is to question taken-for-granted/commonplace assumptions about childhood. The third approach is the structural sociology of childhood:

> *in which childhood is understood to be a permanent social category in society. Its members change, but childhood, in its relations with the other major social group – adulthood – continues as an essential component of a social order where the general understanding is that childhood is a first and separate lifespan whose characteristics are different from the later ones (Mayall, 2002: 23).*

This fits with a life-span, or life-course, approach, which has dominated sociological understandings of childhood as a period of socialisation, in which childhood is understood as a precursor to adulthood. Socialisation is (deliberately) downplayed in the sociology of childhood, in order to focus on the here-and-now of children's (current) lifeworlds. For sociology, the task 'is to link the empirical manifestations of childhood at the level of children's lives with their macro-level contexts, and to focus on the social structures and mechanisms as they may be found to "determine" these manifestations and in this sense help to explain them' (Alanen, 2001: 13). Alanen utilises the concept of generation, and suggests that this distinguishes children as a separate social group and can help to identify the 'structures from which children's powers (or lack of them) derive: the source of their agency... is to be found in the social organisation of generational relations' (Alanen, 2001: 21). As Mayall notes, 'much of the major work in this strand has been done through large-scale work where such major movements as industrialisation, urbanisation, scholarization have been analysed in relation to distributive justice, concepts of work, and the character of children's actual everyday lives' (Mayall, 2002: 23).

Michael Freeman, editor of the *International Journal of Children's Rights*, called for a dialogue between children's rights advocates and sociologists of childhood in a paper published in 1998 (Freeman, 1998). Mayall (2000: 243) explored in more depth the links between the sociology of childhood and children's rights, and suggested that we 'need to extricate children conceptually from parents, families, professionals ... to study the social condition of childhood and write children into the script of the social order'. She emphasises that:

> *Childhood is a political issue. Theories about what children need, about how they develop and what input from adults is appropriate, are indeed theories or stories (rather than facts) and practices that derive exclusively from adult perspectives. They*

derive from adults' study of children, contextualised and structured by adults' social and economic goals in specific societies. Yet in the name of "scientific" formulas about child development and children's needs we need to separate childhood off from politics (2000: 244–5).

Mayall addresses the question of rights:

...in defining childhood as inferior, as objects of adult socialisation – we depersonalise children. In proposing that we know best the best interests of the child, we deny children's rights. We deny them the right to participate in the structuring of their childhoods. Though we may work to protect children and provide for them, we find it much harder to take children seriously as contributors to social thinking and social policies (2000: 245).

Mayall argues that 'the sociological project is to work initially on the task of extracting children theoretically from the family in order to study their social positioning as a social group. A next step is to replace children in reciprocal relations with adults, and childhood with adulthood...' (2000: 247). Following Alanen, Mayall argues that a generational approach 'allows us to recognise that child-adult relations take place between groups of people subject to differing constellations of social, historical and political ideas' (2000: 251).

In the mid-1990s, the social anthropologist Sharon Stephens, in her wide-ranging work on childhood in relation to environmental and political issues (Stephens, 1995), posed important questions about the place of childhood in the political economy. She emphasised the need to recognise the limits of understanding childhood as (merely) a social construction, and pointed out that the consequences of various policy constructions of childhood are very real in their outcomes as they are experienced in everyday life (Ennew & Morrow, 2002: 16). In other words, 'children live real childhoods rather than social constructs' (Ennew & Morrow, 2002: 15). Sociologists of childhood have (perhaps) been wary of researching explicitly the "problems" of "poor" children, not least because of the ethical questions that such research raises, as well as the need to explore and understand the experiences of "ordinary" children in the first place.

2. CHILD POVERTY

The study of child poverty (and poverty in general) has, not surprisingly, been dominated by economists and social policy specialists mainly working within human capital frameworks (see, for example, Bradshaw, 2007; Barnes et al., 2008). Economists define measures of poverty, advise governments, and are the dominant voices in discussions of child poverty in the UK (and elsewhere).

Children are subsumed under the umbrella of "family", meaning parents, and measures are based on parents' income. The dominant policy recommendations are inevitably that parental employment is the solution to children escaping persistent poverty and, certainly in the UK, a huge policy effort has been directed at encouraging parents (mothers as well as fathers) into work. This is despite the fact that the largest group of children in poverty is those with working parents on low wages, not those with out-of-work parents (TUC, 2008). There is also awareness that policies have failed to alleviate the situation for families "with the most entrenched problems" (see, for example, Barnes et al., 2008) and even before the current financial crisis the UK government had begun to acknowledge that it would not meet the ambitious targets it has set itself to reduce child poverty.

Alderson (2008a: 80) has pointed out that there is a tendency in child poverty research inadvertently to support neo-liberal economic approaches and 'to assume that increasing parental employment is the major solution to child poverty, and to confine economics into terms of wealth and poverty, earning power or potential and income'. This kind of research tends to be funded by the World Bank, OECD and national agencies to try to provide global and national policy-makers with answers to questions based upon liberal thinking. An alternative approach, based on the structural sociology of childhood, would attempt to uncover the mechanisms of neo-liberal policies that fail to alleviate (or indeed generate) child poverty. Alderson suggests that children are implicitly treated as objects and "products" of the system. Children are perceived in terms of inputs and outputs, or units of human and economic capital if they are not invisible and ignored in economic policies' (2008a: 81). Alderson (and others, see Bourdieu et al., 1999; Wacquant, 2009) is critical of dominant strands of "poverty" research as potentially stigmatising, 'because these tend to ask people about their difference, isolation, and deficits in their lives and in themselves' (2008a: 86).

The unintended consequences of policies aimed at alleviating child poverty in the UK have been explored by Davies (2008) in a recent study for the Joseph Rowntree Foundation. He describes "povertyism", in other words, the discrimination experienced by families living in poverty. His research was conducted with and by adults who are recipients of various government benefits that are intended to address child poverty. He found that:

> *The use of language with negative connotations to refer to people living in poverty can lead them to greater alienation from society… The emphasis on responsibilities before rights has led to an erosion of basic rights to social protection… Families living in poverty often experience enormous difficulties in accessing their rights to services. They also face povertyism in the form of judgements from other people… Many disadvantaged families live with the fear of their children being taken into care due to the intervention of local authority social services (Davies, 2008: 3).*

Disadvantaged families may be encouraged to place their children in day care, or out-of-school care, but may be reluctant to do so because of a (perhaps understandable) lack of trust (see Barker et al., 2003).[1] Some authors report that children are being removed from their families into public care because their families are too poor to care for them (see Vandenhole chapter 2). This fits with a long history of children being removed to 'orphanages' not because they were orphans but because their parents were too poor to raise them, not least to deter families from seeking any parish help at all (Hendrick, 2003). What is missing from Davies' discussion is any structural analysis of *why* these phenomena are happening in rich countries. It is of prime importance to distinguish what is done in practice in the name of anti-poverty policies and government cutbacks in welfare spending, and what is done in the name of children's rights as enshrined in the UN CRC. There is a real danger of confusing the two approaches, and claiming that children are taken into care as a way of respecting their rights, when in fact the UN CRC could potentially be used to reinforce children's claims to rights to privacy and family life.

Further, from sociological research there is evidence that children (like adults) are acutely aware of the label "poor" and ask researchers not to use it when reporting the results of the research – they find it stigmatising (see e.g. Sime, 2008). Alderson suggests that children in poverty research 'are seldom asked about their views of their authentic identity, values and ways of life, enjoyment and solidarity in their friendships and communities' (Alderson, 2008a: 86), though recent studies have begun to engage with children explicitly to explore their accounts and experiences of living in situations of poverty, disadvantage and social exclusion, and there is now a good body of evidence (see Redmond, 2008a and Attree, 2004 for reviews, as well as Ridge, 2007; Hill et al., 2006; van der Hoek, 2005).

3. THE UN CRC AND CHILD POVERTY

While the sociology of childhood has (necessarily) extricated children from the study of families, the UN CRC firmly locates children within their families, first and foremost.[2] The preamble to the UN CRC states:

> *Convinced that the family, as the fundamental group of society and the natural environment for the growth and well-being of all its members and particularly children, should be afforded the necessary protection and assistance so that it can fully assume its responsibilities within the community (UN CRC Preamble).*

---

[1] Made at the Antwerp meeting by Swa Schijvens, the "expert insider" about "poverty experience".

[2] Indeed, many criticisms of the implementation of the UN CRC in Western developed countries relate to the simplistic idea that children's rights and parents' rights are in direct opposition to each other – this is very misleading, but important to address.

The UN CRC does not contain specific rights relating to child poverty and does not define the term. (Nor would one expect it to, given the difficulties of defining such a contested and contextual concept.) As others have noted (see e.g. Olk & Wintersberger, 2007 for a useful overview, also Fernandes, 2007, and Redmond, 2008b for analysis of UN CRC articles relevant to child poverty), the UN CRC outlines general rights in relation to children's well-being that *relate to* poverty and deprivation. Thus, Article 27 para. 1 stipulates 'the right of every child to a standard of living adequate for the child's physical, mental, spiritual, moral and social development'. Article 27 para. 2 stipulates that parents (or others responsible for the child) 'have the primary responsibility to secure, within their abilities and financial capacities, the conditions of living necessary for the child's development. Article 27 para. 3 states that States Parties are invited 'in accordance with national conditions and within means, to take all appropriate measures to assist parents… to implement this right and… in case of need (to) provide material assistance and support programmes, particularly with regard to nutrition, clothing and housing'. On the other hand, Article 26 regulates children's access to social security: para. 1 states that governments 'shall recognise for every child the right to benefit from social security, including social insurance'; and para. 2 states that 'the benefits should, where appropriate, be granted, taking into account the resources and the circumstances of the child and persons having responsibility for the maintenance of the child…'. As Olk & Wintersberger (2007: 66) note, 'while the task of safeguarding a decent standard of living is seen as the main responsibility of parents, safeguarding the access to social security remains a primary obligation of government'.

Rather, the UN CRC sets out practical measures and, as Redmond notes, 'definitions [of poverty] have to come from elsewhere' (Redmond, 2008b: 64). The articles of the UN CRC (like all international human rights conventions) are deliberately left open to interpretation. Linkages between research evidence and implications for children's rights can usefully be made.

Alderson (2008b: 15) suggests that 'we need to be careful not to confuse rights with wants and demands. UN CRC sets possible minimum standards, which can be enforced where countries have incorporated the UN CRC into domestic law'. As she notes, 'some rights are aspirational, not yet fully realisable', but only:

> *To the maximum extent of (each nation's) available resources. Rights are conditional, affected by the evolving capacities of the child, and responsibilities, rights and duties of parents and national law. Rights cannot be exercised in ways that would harm other people and the child concerned. In exercising rights, people must respect the rights and reputation of others, as well as national security and public order, health, and morals (2008b: 15).*

She notes further:

> *Rights are shared, being about solidarity, equality in social justice… The UN CRC is not about selfish individualism. To claim a right acknowledges that everyone has an equal claim to it and so reaffirms the worth and dignity of every person. … The UN CRC is about broad principles, which can be interpreted and applied in different ways according to local values and traditions. There is therefore at times confusion and disagreement about how best to honour certain children's rights…* (2008b: 16).

In a discussion paper entitled "A child rights approach to child poverty" for Eurochild, Fernandes (2007) makes suggestions, for example, about how a human rights perspective sees children as units of analysis in their own right, and focuses the analysis on a broader concept of resources. These points match well with a sociological approach. An approach based on the UN CRC 'views children as individuals that act independently of their parents' (2007: 3). The point about human rights approaches is that they emphasise the accountability of policy makers and other actors – thus emphasising the role of states. The question here is how to target resources directly to children, rather than through their parents, by improving their environments (broadly defined – to include schools, public spaces and so on). Where the CRC is incorporated into domestic law, a rights-based approach can enable children to claim their rights legally. In the UK, on the other hand, where the CRC has yet to be incorporated, children have recourse to the Human Rights Act. There have been some examples of cases brought to the European Court of Human Rights (Lyon, 2007), but none as yet related to social rights, or to economically-linked welfare rights, related to services (such as child benefit, free health services, education, services for disabled children, or social services). As Fernandes suggests: 'The human rights approach … takes us beyond the idea of child poverty centred on need. It puts less emphasis on individuals' personal failure to justify child poverty, focusing more on the failure of macro-economic structures and policies, which is a responsibility of the state and other national and international institutions' (2007: 5). In summary, the UN CRC has provisions that are relevant to the study of, and policy responses to, child poverty. The next section discusses social exclusion and human capabilities approaches to child poverty.

4. SOCIAL EXCLUSION AND HUMAN CAPABILITIES APPROACHES: LINKING TO THE UN CRC

In a recent paper, Redmond (2008b) has evaluated economic welfare, capabilities and social exclusion approaches to child poverty, and explores what an approach based on the UN CRC would look like. Income-based economic welfare approaches use money as an indicator, which is meaningful to policy-makers, but is it less relevant to children? Social exclusion came into use during the 1980s alongside the language of poverty, and it emphasises the processes and practices 'by which individuals or households experience deprivation, either of resources (such as income), or of social links to the wider community or society' (Scott & Marshall, 2005: 204; see Room, 1999).

Human capabilities approaches derive from Amartya Sen's work on Human Development (1999) and emphasise the 'capabilities that a person has, that is, the substantive freedoms he or she enjoys to lead the kind of life he or she has reason to value' (Sen, 1999: 87). For Sen, 'poverty must be seen as the deprivation of basic capabilities rather than the lowness of incomes' (1999: 87). Sen acknowledges that 'inadequate income is a strong predisposing condition for an impoverished life' (1999: 87), but suggests capabilities approaches can focus on deprivations that are intrinsically important (instead of instrumentally important); that there are other important influences on capability deprivation; and thirdly, that there is a relationship between low income and low capabilities that can be treated as a variable – and can be affected by 'age, gender and social roles, location, and so on' (1999: 88). Sen is not particularly interested in children and childhood in a sociological sense, but as Redmond suggests, the capabilities approach offers a vision of what freedom from poverty should mean and involves the development of indicators that might describe different aspects of children's moral, spiritual, mental and social well-being or development – the problem, Redmond notes, is translating these into rights.

Social exclusion approaches on the other hand, as Redmond suggests, focus on processes as well as outcomes, and he suggests a synthesis between human capabilities and social exclusion approaches, to balance economic welfare approaches. He also identifies a number of challenges to a recognisable rights-based definition of poverty:

> 1: A rights-based definition might draw heavily on concepts integral to capabilities and social exclusion approaches, but these will not be rigorous enough to be acceptable to economic welfare approaches (2008b: 78). In other words, there would need to be conversion factors that would translate outcomes based on rights.
> 2: Both the capabilities and social exclusion approaches offer considerable guidance as to how thresholds of capabilities or inclusion should be determined (2008b: 79).

Redmond argues that 'a synthesis of the two approaches would suggest that acceptable thresholds for children's "being" and "becoming" encompass both the optimality that is implicit in the capabilities approach (and in the CRC) and the inequality aversion that is inherent in the social exclusion approach' (Redmond, 2008b: 79). He suggests that '[a] rights-based poverty definition and threshold should act as a guide for policymakers at national level to ensure redistribution of resources (something that governments can directly control) towards equitable outcomes (something that governments should aim to achieve, but over which they have perhaps less direct control). … monitoring would be necessary…' (2008b: 79). He concludes:

> A universal right-based definition of child poverty should encompass indicators of children's "being" and "becoming" that are … based on an agreed upon position on minimum levels of equity that would ensure that all children are included and have the opportunity to lead lives they have reason to value (2008b: 79).

However, in my view it is not clear precisely what human capabilities approaches add to this debate. Internationally agreed conventions (like the UN CRC) *can* be translated into tools and indicators (Ennew & Miljiteig, 1996). There seems to be considerable slippage between rights discourses and economic concepts, and this begs the question of why economists are so uneasy about rights-talk. Nussbaum, confusingly, who has worked with Sen on a list of ten Central Human Capabilities, suggests that the capabilities approach, as she has developed it, 'is a species of a human rights approach' (2007: 21). Ultimately, however, some have argued (from a Marxist viewpoint) that human capabilities approaches revert easily and quickly to straightforward human capital approaches (see Dean, 2009 for a stringent critique). Human capabilities approaches are ultimately individualistic and instrumental/functionalist, and arguably have little to offer our thinking about children. As Redmond notes, capabilities approaches (like economic welfare approaches) offer 'little space for agencies of duty-bearers with respect to groups of people' (2008b: 70). Dean further points out that capabilities approaches do not 'challenge the roots of social injustice or the quotidian relations of power' (2009: 274). A view from the sociology of childhood would focus on children's present lived experiences, instead of a focus on what they have the potential to become, which the human capabilities approach suggests.

A final point is that the links between children's well-being and poverty are still not well elaborated or understood, and it is not straightforward to trace causal pathways – setting aside material well-being, it is not possible to demonstrate that poor children are necessarily the "unhappiest" or suffer the most "ill-being". For example, there is (developmental psychological) research evidence from the USA that suggests that children of affluent parents are at a greater "risk" of various "non-trivial threats" to their psychological well-being than their poorer counterparts, risks such as higher rates of substance abuse, anxiety and depression that are linked to two sets of factors: excessive pressure to achieve, and physical and emotional isolation from parents (Luthar, 2003; Luthar & Latendresse, 2005). Conversely, some recent research has found that the children in the UK who are the "happiest" are living in some of the most deprived parts of the UK (OFSTED, 2007). "Happiness" research in relation to children or adults is profoundly problematic, however, and is probably not useful for the purposes of this discussion (see Morrow & Mayall, 2009 and Suissa, 2008 for a critique of "positive psychology"). Studies explicitly focused on children's experiences of inequality and relative deprivation could add to the debates here.

As noted, a good deal of research has now been carried out that is based on children's accounts of their experiences of poverty. However, policy recommendations persist in suggesting that improving family incomes through parental work and providing childcare will automatically improve the situation for children. In the UK context (and elsewhere), benefits are increasingly linked to conditions; in other words, 'people experiencing poverty are increasingly expected to meet ever more demanding responsibilities in order to meet the criteria for claiming entitlements' (Davies, 2008: 6). This has the effect of denying people their basic human rights, as Davies notes, and he argues that attitudes amongst service providers and policy-makers need to change. He suggests better awareness and understanding of the powerlessness that benefit recipients feel, together with a rights-based perspective, are possible solutions.

Sandbaek (2007: 194), in relation to Norway, argues that policies should be directed to children themselves, based upon their rights as citizens. Her argument for a rights-based approach 'is that children have independent claims on society. Society should not longer accept that children grow up in poverty. ... a rights-based approach puts the role of the state on the agenda'. She does warn, however, that we must not presuppose children are 'autonomous, unsupported individuals'. Children themselves emphasise their connectedness to others rather than their autonomy and separateness (see e.g. Morrow, 2001b; Mayall, 2002). This is not to imply that the opportunities for children to realise their rights, to access services, to overcome thresholds, depends on their "innate" strength and is not dependent on their education and other forms of capital, but rather to suggest that (adults) need to recognise the constraints upon children's agency that may be influenced by their experiences of social exclusion (which in turn may relate to poverty).

Sandbaek suggests that we need a 'concept of children as agents that includes their right to protection, dependency and care' (2007: 195). Another pitfall is 'to underestimate children and approach them with an idealised perception of what childhood *should be* like' (2007: 195, emphasis added). Further, '[i]t is by no means obvious that every kind of restriction on children's consumption or activities should be regarded as negative' (2007: 195). She suggests that:

> *A rights-based approach in order to secure children's needs should also not be practised in a way that creates contradictions between parents and children. On the contrary,*

there is reason to build on parents' efforts and willingness to take care of and give priority to children. At the same time it is also necessary to recognise that these expectations are putting poor parents under considerable pressure. ... It is likely to be perceived as a positive contribution if the state shares the responsibility for safeguarding the rights to a decent standard of living of the youngest generation, while at the same time recognising the efforts made by most parents to bring up their children (2007: 195).

The precise role of the state of course differs between political systems, and it is no surprise that poverty and inequality are more acute in conservative countries (US, UK) than in social democratic (Scandinavian) countries (Esping Anderson, 1990).

5. SOCIAL EXCLUSION (AND SOCIAL CAPITAL)

An example of research within the new sociology of childhood paradigm by the author that explores processes of social exclusion illustrates some of the points raised above. In the late 1990s, Morrow (1999a) explored the usefulness of the slippery concept of "social capital" for thinking about children and young people's health and well-being in "deprived areas" (though this was not directly conceived of as "poverty research").[3] "Social capital" consisted of the following features: social and community networks; civic engagement or participation; community identity and sense of belonging; and norms of co-operation, reciprocity and trust in others within the community (Putnam, 1993). The premise was that levels of "social capital" in a community have an important effect on people's well-being. "Social capital" is a concept that has been contested at a number of levels, conceptually, methodologically, and theoretically (discussed elsewhere, see Morrow 1999a, 2001a, b). The study explored what these elements of social capital might mean for children and young people, and prioritised their accounts.

The research was conducted in two schools in relatively deprived wards in a town in SE England (disguised as "Springtown"; children chose their own pseudonyms; the site was chosen to match another HEA study on adults and social capital). One ward (West Ward) consisted of "suburban sprawl", with post-war housing and factories; the second (Hill Ward) consisted of a mixture of industrial development and Victorian, inter-war and post-war housing development. The sample comprised 101 boys and girls in two age bands: 12–13 year olds and 14–15 year olds, with a proportion from minority ethnic groups. The study used a combination of qualitative research methods, including visual methods and

---

[3] Funded by the Health Education Authority, then the health promotion arm of the UK Government's Department of Health.

group discussions, and structured activities to explore young people's subjective experiences of their neighbourhoods, the nature of their social networks, and their experiences of participation in decision-making in schools and neighbourhoods (for full details, see Morrow, 2001a).

The centrality of friendship to the everyday lives of children and young people was very clear. Proximity to friends often seemed to determine how children felt about where they lived. Friends were central to many activities outside school, and were also important sources of emotional support. "Being part of the group" was clearly crucial; one discussion with a group of 15–16 year olds focused on peer pressure (one boy used this term and the others clearly understood it) and Amy explained what it meant:

Amy: it's blending in with the rest of the group, if the rest of the group is wearing Nike trainers, you feel like you've gotta have Nike trainers, if the rest of the group is smoking, you feel like you've gotta smoke.
Gizmo: even with Nike trainers, as well, it's like, oh, they're the old model, this is the new model, you must have this, and you haven't really got any free choice to wear what you want.

The groups and associations that matter to young people may consist of clear hierarchies. Amy described it thus:

You get class peer groups, you get like first class, like they're all popular, and if like say a third class person walked past 'em, they'd be jeering. They're exactly the same age, they know them, but in the street… you get some people who are so popular, everyone will be, often they're really horrible people as well, but everyone's like 'oh, yeah, let's join their gang', then you get these other people, that nobody seems to like, and they're really nice people, and they walk past the other people, and they're like jeering at them, they've done nothing wrong.
Dave said: that's back to clothes, and stuff, again, though, innit.

In School 2, in a group discussion of younger children, the conversation turned to shopping and clothes when Henry said 'if you don't have all the stuff, you're not like, *with* everyone'. Paris, another boy, explained: 'Miss, you have to have fashionable clothes, or you ain't wiv us'. Ricardo said 'I asked my mum for, like, designer clothes, she said it's just the same clothes but just with a little label on them', to which the other children chorused 'Yeah, it's true!' Carlos responded 'but you just get tooken the mick [i.e. made fun of] out of if you don't have them'.

These examples illustrated some interrelations between social capital and symbolic capital. Belonging to a particular "class peer group" (or being "with us") by having access to the appropriate symbolic markers may provide a sense of belonging that relates to well-being, but (paradoxically) may also set up habits (in this case, smoking) that may be damaging for future health. As Amy said, 'you feel like you've gotta smoke'. The economic capital in this example was veiled – the "right" trainers were expensive, but which ones were defined as the "right" ones is a matter of symbols (signifiers). Further, the financial expense of the trainers was also likely to be part of the symbolism.

Negative symbolic capital related to children's low social status – the simple fact of "being young" – could also be seen in young people's descriptions of relationships in their town and neighbourhoods. These relationships were characterised by a perceived lack of trust from the adults around them, and they were well aware of how they were perceived negatively, particularly in the town centre. There was one very lengthy discussion that hinged around the issue of being regarded with suspicion:

> *Amy: it's horrible, 'cos you walk into a shop, you've got no bag on you, you're looking quite smart, and you've got all these security guards watching you like a hawk, /.../*
> *Olanda: they stereotype us. /.../*
> *Dave: but then again, then again, even though they do do that to us, how many people do you know that steal things, /.../*
> *Amy: but it's the few, it's the little minority of people nicking it that's letting us down, it's not nice going into a shop and you've got this big security guard watching you, you feel uncomfortable, you've gotta walk out of the shop, so not only are they losing business, but they're also losing our respect as well, 'cos I don't wanna go back in the shop.*
> *Gizmo: the security guards think, 'oh my god here come the 15 year olds. Let's watch them.' Then if somebody else comes in steals something and runs out, they're going like 'oh we'll leave him, we'll just concentrate on the 15 year olds'. You know.*

In a discussion with 15-16 year olds, one boy said news media representation of young people was "insulting". This led the others to say:

> *Isabelle: 'cos people, like, expect you to be no good, or like, alcoholics*
> *Maria: stereotypes*
> *Maggie: or like drug dealers, smokers*
> *Maria: they don't give you the benefit of the doubt, they just write you off before they've even met you...*
> *Isabelle: just because of your age, so like, they look down on you, and assume that you're causing trouble... when really you're not.*

On news media imagery of their town and neighbourhoods, they said:

Melissa: it makes us feel bad about our community, and where we live, and everything.
Boy: It makes you feel bad, because it just shows the bad things in the newspaper, it don't show the good things, you get the bad things. /.../
Maggie: well, bad things come in the newspaper, and people think, 'oh, I don't really wanna live there, it's such a bad place'.
Boy: they just see what's on the news and think that's it.
Melissa: most people over-exaggerate, sometimes, don't they?

The study highlighted how "children" and "young people" are not homogeneous categories, but revealed a range of social identities. For girls, personal safety was a crucial issue, and sexual assault was perceived as a threat. There were differences in children's accounts according to age: the expressed needs of 12 year olds, for example, for places to play and make dens, differed from the expressed needs of 15 year olds, for example for places to socialise away from the sometimes hostile gaze of adults. Further, there were differences in priorities according to ethnic background, and racial harassment was described by several children. Clearly, this harassment is likely to lead to social and emotional exclusion.

The study also illustrated how this age group feels a range of practical, environmental and economic constraints. These included not having safe spaces to play in or hang out in, not being able to cross the road because of the traffic, having no place to go except the shopping centre, but being regarded with suspicion because of lack of money. The extent to which children and young people move around freely to participate in activities with their friends is likely to be constrained by the physical geography of the built environment, issues of community safety and traffic, parental norms about when children may go out, as well as access to financial resources. The accounts from the children revealed the extent to which their lives were constrained by material circumstances and structural factors at the macro-level that were located beyond their spheres of influence. Further, arguments about intangible and ill-defined notions like "trust", "community engagement" and "reciprocity" may mask more tangible inequalities along the lines of poverty and deprivation.

The UN CRC was utilised to analyse these data, and the final report suggested that social policy should pay attention to children's quality of life, in the broadest sense, in the here and now, rather than be driven by a perspective which prioritises children as future citizens, in terms of human capital. Young people were well aware that they were effectively denied a range of participatory rights that adults take for granted. They were not as rebellious and disaffected as dominant imagery depicts them to be. They had a strong sense that they needed their educational qualifications, and that school was important. At the same time, they wanted to have access to safe local streets and neighbourhood spaces, but they were well aware that their needs were neglected. The UK Government remains ambivalent about the UN CRC (Hendrick, 2003). As Freeman has suggested, we should see the Convention as a beginning, 'but the lives of children will not change for the better until the obligations it lays down are taken seriously by legislatures, governments and all others concerned with the daily lives of children' (Freeman, 2002: 115).

Over ten years later, the same criticism can be made: Children's Rights Alliance for England CRAE (2008), in their alternative report to the Committee on the Rights of the Child, argue that there is 'a profound lack of engagement with the spirit and requirements of the UN CRC by ministers and officials in England… we still do not have a government that embraces the Convention as a tool to transform children's lives'[4] (CRAE, 2008: 1). The Committee on the Rights of the Child noted in its concluding observations relating to the UK (2008) that:

> *The committee recommends that the State party… allocate the maximum extent of available resources for the implementation of children's rights, with a special focus on eradicating poverty and reduce inequalities across all jurisdictions… (UN Committee on the Rights of the Child, 2008: 5).*

6. CHILDREN'S PARTICIPATION

One of the four general principles of the UN Convention on the Rights of the Child is that of participation. Article 12 stipulates that 'States parties shall assure to the child who is capable of forming his or her own views the right to express those views freely in all matters affecting the child, the views of the child being given due weight in accordance with the age and maturity of the child…'. This has led to a wave of research on children's participation (see, for example, Percy-Smith & Thomas, 2009) and a greater understanding of the possibilities and limitations of children's participation. It has become increasingly apparent that children are effective, resourceful commentators on their situations, when asked. A much greater challenge is to get adults to act upon what they hear. Further, children's participation needs to be understood in the context of adult's participation (or lack of it, particularly in situations of deprivation) (explored in Morrow, 2008). However, there is a tendency among children's rights advocates

---

[4] It should be noted that approaches to the UN CRC differ between the devolved UK governments, and Northern Ireland, Scotland and Wales have embraced the UN CRC more fully than the English government.

to make claims about children's participation that are overly optimistic. For example, Fernandes (2007: 9) suggests:

> *The EC should develop practical proposals to ensure respect for children's right to be heard in all matters concerning them. Children's participation must be systematic, structured and adequately resourced. Most importantly, children's opinions, particularly the voices of most vulnerable children, should be listened to and taken seriously in the development of effective policies and strategies to prevent and eradicate child poverty.*

This is a big leap, and places a burden upon children to reflect upon their own situations and find solutions for what are problems of politics and economics. Further, it runs the risk of assuming a kind of pathology of child poverty. Similar arguments have been made about adults' participation in poverty research and action in developing countries (Cooke & Kothari, 2001). One can find repeated examples where children's views of what matters to them do not fit with adult-driven agendas. Children and young people do not necessarily define themselves as "poor" and may have other, more pressing things to deal with. On the other hand, the concerns that children have about their everyday lives may be implicitly linked with poverty as an underlying cause, in ways that they do not explicitly express or articulate in research. Children tend to express great pride in their parents, and particularly their mothers, and are acutely aware of the ways their parents may struggle to provide for them (e.g. Morrow, 1998, 2001a; Ridge, 2006). As noted, children also tend to express themselves relationally, emphasising their interdependence with other family members and their friends. Children and young people who experience poverty may have important views about their experiences, and these views need to be respected too. The question is how to undertake research that is sensitive to the pitfalls of the negative connotations of the term "poverty" and that does not further stigmatise disadvantaged groups of children (Sime, 2008).

Research has tried to uncover what children mean by the term "well-being" (another slippery concept that has recently entered policy debates about childhood and poverty), from their viewpoints. In Australia, Fattore et al. (2007) undertook qualitative research with a sample of 126 rural and urban 8–15 year olds from a range of socio-economic backgrounds in New South Wales, (the age and gender breakdown of the sample are given, but not the ethnic background, so it is not possible to determine whether aboriginal children were included) to explore children's definitions of "well-being" and found that:

> *Well-being is defined through feelings, in particular happiness, but integrating sadness is also relevant. Well-being is about feeling secure, particularly in social relations… also as being a moral actor in relation to oneself (when making decisions in one's best*

> *interest) and when one behaves well towards others… Well-being is the capacity to act freely and to make choices and exert influence in everyday situations. This was not necessarily being independent from others. Children articulated the social relations upon which autonomy was premised, including stable, secure relationships with adults… (Fattore et al., 2007: 18).*

They also found that children articulated clear ideas about the importance of having a positive sense of themselves, material resources, physical environment and home. Qualitative studies show great potential for complementing, or expanding upon, larger-scale survey research by providing deeper insights into children's everyday lives and what matters to them from their point of view. However, as noted earlier, underlying all this must be attention to precisely how economic deprivation/poverty affects families and, thus, children directly, and to incorporate attention to political-economic processes and structures that create, or fail to reduce, poverty in the first place.

Finally, some writers are calling for a revision of economics. Alderson (2008a: 87) draws on critical realism and green economics to suggest that these can help to analyse:

> *How the hope of relative wealth draws children into every-rising moral expectations about what a good parent should provide, leading to higher consumption, longer working hours, higher debts, and ever-rising levels of comparative wealth. There are always richer children to catch up with, though whether they are happier is seldom investigated.*

As Alderson suggests, what sociologists can do is to question the context (2008a: 87): 'the experience of relative poverty may be more psychological and relational than material'. However, this begs some important questions, because consumption and the market are only one part of the question about the effects of poverty on children. While poverty, like childhood, may be socially constructed (in research), the consequences of it are lived, embodied and experienced and felt by children in subtle or acute ways. Ridge (2007) is careful to suggest that the social implications of poverty are as or more important than the material consequences. She teases out the links between material deprivation and its social consequences for children's relationships, belonging and inclusion.

7. CONCLUSION

This chapter has suggested that sociological approaches to the study of childhood could be utilised, firstly, to explore children's experiences of poverty and deprivation; secondly, to link these experiences to rights as enshrined in the UN

CRC; thirdly, to try to help explain why there are barriers to implementing children's rights in specific instances; and finally, to help make links between the two dominant discourses: that of child poverty (based on economics, social exclusion and human capabilities) on the one hand, and children's rights on the other. However, the sociology of childhood remains a marginal critique, certainly in the UK policy context, in the face of huge structural pressures and the relentless emphasis on what are ultimately human capital approaches based on "outcomes" and realising children's potential as productive, hard-working adults.

Sociological analysis (whether childhood sociology, or sociology more generally) has the potential to illuminate some of the processes that operate to encourage social exclusion and discrimination related to poverty. As Freeman noted over ten years ago, sociology '… can throw light on the wrongs from which children suffer and may help us to understand why these wrongs are perpetuated against children' (Freeman, 1998: 442). In relation to rights, a sociological lens can be useful to bring to bear as an analytic tool to explain why it is so difficult to realise rights in practice. The challenge remains to get policy makers to set aside the dominant (economic) vision of childhood as entirely based on futurity (what children will become), to listen, and understand what childhood is like now, and to change their policies and practices accordingly.

ACKNOWLEDGEMENTS

The author would like to thank the seminar participants, Berry Mayall and Leena Alanen, and the editors for their helpful suggestions on an earlier version of this chapter. She would also like to thank the schoolchildren who participated in the research reported here.

REFERENCES

ALANEN, L. (2001), Explorations in generational analysis, in: ALANEN, L. & MAYALL, B. (eds.), *Conceptualising child-adult relations.* London: Routledge.

ALDERSON, P. (2008a), Economic alternatives and childhood poverty, *International Journal of Green Economics,* (2): 1, 77–94.

ALDERSON, P. (2008b), Youth people's rights: children's rights or adults' rights? *Youth & Policy,* (100), 15–25.

ATTREE, P. (2004), Growing up in disadvantage: a systematic review of the qualitative evidence, *Child: Care, Health and Development,* (30): 6, 679–689.

BARKER, J., SMITH, F., MORROW, V. et al. (2003), *The impact of out of school care on children and families*. London: DfES. www.dfes.gov.uk/research/data/uploadfiles/RR446.pdf

BARNES, M., CONOLLY, A. & TOMASZEWSKI, W. (2008), *The lives of children in persistently poor families. Research findings*. London: NatCen.

BOURDIEU, P. et al. (1999), *The weight of the world. Social suffering in contemporary society*. Cambridge: Polity Press.

BRADSHAW, J. (2007), Some problems in the international comparison of child income poverty, in: WINTERSBERGER, H., ALANEN, L., OLK, T. et al. (eds.), *Childhood, generational order and the welfare state: exploring children's social and economic welfare*. Vol. 1, COST A19: Children's welfare. Odense: University Press of Southern Denmark.

CRAE (Children's Rights Alliance for England) (2009), *State of children's rights in England report 2009*. London: CRAE.

COOKE, B. & KOTHARI, U. (2001), *Participation: the new tyranny?* London: Zed Press.

DAVIES, M. (2008), *Eradicating child poverty: the role of key policy areas. The effects of discrimination on families in the fight to end child poverty*. Joseph Rowntree Foundation, www.jrf.org.uk, accessed 2 September 2009.

DEAN, H. (2007), Social policy and human rights: rethinking the engagement, *Social Policy & Society*, (7): 1, 1–12.

DEAN, H. (2009), Critiquing capabilities: the distractions of a beguiling concept, *Critical Social Policy*, (29): 2, 261–278.

ENNEW, J. & MILJETEIG, P. (1996), Indicators for children's rights: progress report on a project, *International Journal of Children's Rights*, (4), 213–236.

ENNEW, J. & MORROW, V. (2002), Releasing the energy: celebrating the inspiration of Sharon Stephens. Introduction to Special Issue of *Childhood: A global journal of child research, 'Children and the politics of modernity: a tribute to Sharon Stephens'*, (9): 1, 5–17.

ESPING-ANDERSON, G. (1990), *The three worlds of welfare capitalism*. Cambridge: Polity Press & Princeton: Princeton University Press.

FATTORE, T., MASON, J., WATSON, E. (2007), Children's conceptualisation(s) of their well-being, *Social Indicators Research*, (80), 5–29.

FERNANDES, R. (2007), A child rights approach to child poverty. Discussion paper for Eurochild. www.eurochild.org, accessed 4/3/2010.

FREEMAN, M. (2002), Children's rights ten years after ratification, in FRANKLIN, B. (ed.), *The new handbook of children's rights. Comparative policy and practice*. London: Routledge.

FREEMAN, M. (1998), The sociology of childhood and children's rights, *International Journal of Children's Rights*, (6), 433–444.

HENDRICK, H. (1990), Constructions and reconstructions of British childhood: an interpretive survey, 1800 to the present, in: JAMES, A. & PROUT, A. (eds.), *Constructing and reconstructing childhood: contemporary issues in the sociological study of childhood*. London: Falmer.

HENDRICK, H. (2003), *Child welfare. Historical dimensions, contemporary debate*, Bristol: Policy Press.

HILL, M., TURNER, K., WALKER, M., et al. (2006), Children's perspectives on social exclusion and resilience in disadvantaged urban communities, in: TISDALL, K., DAVIS, J., HILL, M. & PROUT, A. (eds.), *Children, young people and social inclusion: participation for what?* Bristol: Policy Press.

LUTHAR, S. (2003), The culture of affluence: psychological costs of material wealth, *Child Development*, (74): 6, 1581–1593.

LUTHAR, S. & LATENDRESSE, S. (2005), Children of the affluent. Challenges to well-being, *Current Directions in Psychological Science*, (14): 1, 49–52.

LYON, C.M. (2007), Interrogating the concentration on the UN CRC instead of the ECHR in the development of children's rights in England? Policy review, *Children & Society*, (21),147–153.

MAYALL, B. (2000), The sociology of childhood in relation to children's rights, *International Journal of Children's Rights*, (8), 243–259.

MAYALL, B. (2002), *Towards a sociology for childhood. Thinking from children's lives.* London: Open University Press.

MIDDLETON, S., ASHWORTH, K., BRAITHWAITE, I. (1997), *Small fortunes.* JRF/York Publishing Services.

MORROW, V. (1998), *Understanding families: children's perspectives.* London: National Children's Bureau/Joseph Rowntree Foundation.

MORROW, V. (1999), Conceptualising social capital in relation to the well-being of children and young people: a critical review, *The Sociological Review*, (47): 4, 744–765.

MORROW, V. (2001), *Networks and neighbourhoods: children's and young people's perspectives*, Report for the Health Development Agency Social Capital for Health series. London: H.D.A.

MORROW, V. (2001b), Young people's explanations and experiences of social exclusion: retrieving Bourdieu's concept of social capital, *International Journal of Sociology and Social Policy*, (21): 4/5/6, 37–63.

MORROW, V. (2008), Dilemmas in children's participation in England, in: INVERNIZZI, A. & WILLIAMS, J. (eds.), *Children and Citizenship.* London: Sage.

MORROW, V. & MAYALL, B. (2009), What is wrong with children's well-being in the UK? questions of meaning and measurement, *Journal of Social Welfare and Family Law*, (31): 3, 213–225.

MORROW, V. & MAYALL, B. (2010), Measuring children's well-being: some problems and possibilities, in: MORGAN, A., DAVIES, M. & ZIGLIO, E. (eds.), *Health assets in a global context: theory, methods, action.* Amsterdam: Springer.

NUSSBAUM, M. (2007), Human rights and human capabilities, *Harvard Human Rights Journal*, (20), 21–24.

OFSTED (2007), TellUs2 survey results. Press release, www.ofsted.gov.uk, accessed 4.6.2008.

OLK, T. & WINTERSBERGER, H. (2007), Welfare states and generational order, in: WINTERSBERGER, H., ALANEN, L., OLK, T. et al. (eds.), *Childhood, generational order and the welfare state: exploring children's social and economic welfare*, Vol. 1, COST A19: Children's welfare. Odense: University Press of Southern Denmark.

PERCY-SMITH, B. & THOMAS, N. (eds.) (2009), *Handbook of children and young people's participation: perspectives from theory and practice.* London: Routledge.

PROUT, A. & JAMES A. (1990), A new paradigm for the sociology of childhood? provenance, promise and problems, in: JAMES, A. & PROUT, A. (eds.), *Constructing and reconstructing childhood: contemporary issues in the sociological study of childhood*. London: Falmer.

PUTNAM, R. (1993), *Making democracy work. Civic traditions in modern Italy*. Princeton, NJ: Princeton University Press.

QVORTRUP, J. (1987), Introduction, *International Journal of Sociology*, Special issue, 'The sociology of childhood', (17): 3, 3–37.

QVORTRUP, J. (1985), Placing children in the division of labour, in: CLOSE, P. & COLLINS, P. (eds.), *Family and economy in modern society*. Basingstoke: Macmillan.

REDMOND, G. (2008a), *Children's perspectives on economic adversity: a review of the literature*. UNICEF, Innocenti Research Centre. IDP Discussion paper 2008-1.

REDMOND, G. (2008b), Child poverty and child rights: edging towards a definition, *Journal of Children and Poverty*, (14): 1, 63–82.

RIDGE, T. (2002), *Childhood poverty and social exclusion. From a child's perspective*. Bristol: Policy Press.

RIDGE, T. (2006), Childhood poverty: a barrier to social participation and inclusion, in: TISDALL, K., DAVIS, J., HILL, M. & PROUT, A. (eds.), *Children, young people and social inclusion: participation for what?* Bristol: Policy Press.

RIDGE, T. (2007), Negotiating child poverty: children's subjective experiences of life on a low income, in: WINTERSBERGER, H., ALANEN, L., OLK, T. et al. (eds.), *Childhood, generational order and the welfare state: exploring children's social and economic welfare*. Vol. 1, COST A19: Children's welfare. Odense: University Press of Southern Denmark.

ROOM, G. (1999), Social exclusion, solidarity and the challenge of globalisation, *International Journal of Social Welfare*, (8): 3, 166–174.

SANDBAEK, M. (2007) Children's rights to a decent standard of living, in: WINTERSBERGER, H., ALANEN, L., OLK, T. et al. (eds.), *Childhood, generational order and the welfare state: exploring children's social and economic welfare*. Vol. 1, COST A19: Children's welfare. Odense: University Press of Southern Denmark.

SCOTT, J. & MARSHALL, G. (2005), *Dictionary of sociology*. Oxford: OUP.

SEN, A. (1999), *Development as freedom*. Oxford: OUP.

SIME, D. (2008), Ethical and methodological issues in engaging young people living in poverty with participatory research methods. *Children's Geographies*, (6): 1, 63–78.

STEPHENS, S. (1995), Introduction: children and the politics of culture in 'late capitalism', in: STEPHENS, S. (ed.), *Children and the politics of culture*. Princeton, NJ: Princeton University Press.

SUISSA, J. (2008), Lessons from a new science? On teaching happiness in schools, *Journal of Philosophy of Education*, (42): 3–4, 575–590.

TUC COMMISSION ON VULNERABLE EMPLOYMENT (2008), *Hard work, hidden lives: the full report of the TUC Commission on Vulnerable Employment*. London: TUC.

UN COMMITTEE ON THE RIGHTS OF THE CHILD (2008), 49th Session, *Concluding Observations United Kingdom of Great Britain and Northern Ireland*. CRC/C/GBR/Co/4.

UNICEF (2007), Child Poverty in Perspective: an overview of child well-being in rich countries, *Innocenti Report Card No. 7*. Florence, Italy: UNICEF, Innocenti Research Centre. www.unicef.org/irc.

VAN DER HOEK, T. (2005), Through children's eyes: an initial study of children's personal experiences and coping strategies growing up poor in an affluent Netherlands. *Innocenti Working Paper*, 2005–06. Florence, Italy: UNICEF, Innocenti Research Centre.

WACQUANT, L. (2009), *Punishing the poor. The neoliberal government of social insecurity*. Duke University Press, N.C.

4. CHILD POVERTY, CHILDREN'S RIGHTS AND PARTICIPATION: A PERSPECTIVE FROM SOCIAL WORK

Rudi Roose, Griet Roets, Didier Reynaert and
Maria Bouverne-De Bie

INTRODUCTION

Over recent decades, child poverty has remained a stubborn problem in most Western societies, especially in light of the realisation of human rights in general and children's rights in particular (Lister, 1997; Lawy & Biesta, 2006; Cantillon, 2008). The UN Convention on the Rights of the Child (UNCRC) serves as a framework for taking into account the comprehensive and multi-dimensional nature of child poverty (UNICEF, 2000, 2007). Under the terms of the UNCRC, the main argument in international circles is that children's rights are adversely affected by poverty (Walker & Walker, 2002), including civil and political rights but most significantly social rights in such fields as social security, housing, education and health care (Ife & Morley, 2002).[1] Child poverty is explicitly perceived as an existential condition in which there is an increased and fundamental risk of a violation of children's rights (Fox Harding, 1996).

The issue of child poverty is currently high on the policy agenda of the European Union and its member states. At the Lisbon Council meeting in 2000, the EU member states agreed on common objectives for moving towards the prevention,

---

[1] Here we refer to Marshall's articulation of a coherent description and analysis of citizenship (see Lister, 1997; Lawy & Biesta, 2006). Crucially, his typology was grounded within an historical framework involving three elements (see Lawy & Biesta, 2006). The *civil component*, which he traced through legislation that developed largely in the eighteenth century (between 1688 and 1832), includes the rights to freedom of speech, to justice and to own property. *Political rights*, including the right to vote and to stand for political office, followed in the nineteenth and early twentieth centuries, when the franchise was extended to include the majority of the adult population. The final component, *social rights*, which included social security, health care and education, developed mainly in the twentieth century. Each of these three components corresponds to a particular set of institutions – civil rights to the court system, political rights to the institutions of local government and parliament, and social rights to the welfare state.

reduction and elimination of child poverty and social exclusion among children as a specific target (Hoelscher, 2004). Child poverty has featured as a political priority in many national action plans on poverty and social exclusion over the past decade, with parallel developments and concern in Belgium (Steenssens et al., 2008). More recently, there has been a particular concern to show tangible results from the efforts made in light of the European Year for Combating Poverty and Social Exclusion in 2010 (EU, 2007). Amongst the European Year's priorities is, in particular, fighting child poverty, including poverty within families and the inter-generational transmission of poverty (EU, 2008).[2]

In times when child poverty is addressed by social policy makers and civil society as an urgent problem, it is, however, essential to turn a critical eye on underlying problem constructions and on policy and practice interventions following therefrom. Putting the development of anti-poverty policy agendas in a historical and international perspective illuminates the way in which policy constructs its objects of intervention. In framing child poverty as a problem that needs interventions and solutions (Platt, 2005), child poverty is made particularly politically salient under certain ideological conditions, so that it becomes fit for future interventions by policy makers and practitioners in social work and related disciplines (see Jones, 2002). During recent decades, these anti-poverty interventions have been ostensibly undertaken in the name of participation (Lister, 2002, 2004; Craigh, 2003; Bouverne-De Bie, 2005; Vranken et al., 2004, 2008). Nevertheless it is often not clear what participation and the right to participate really mean, partly because the participation discussion remains under-theorised. We argue that participation is not an *inherently* positive and unambiguous notion at all because it mirrors ideological motives of social policy, civil society and practice (Roose, 2005). In this chapter, we relate the question of child poverty and children's rights to participation as a key concept in the debates on combating child poverty in the area of social work (Bouverne-De Bie et al., 2003).

1. IN THE NAME OF PARTICIPATION

Over the past century, child poverty has been the object of interventions by social policy makers and researchers (Platt, 2005), and by various disciplines that inform social work practices with children and parents living in poverty (Jones, 2002). Child poverty is currently attracting increasing interest from disciplines

---

[2] In line with the priority policy areas identified in the *Strategic Framework Document* (EU, 2008), the European Year focuses on paying special attention to large families, single parents and families caring for a dependent person, as well as poverty experienced by children in institutions, and on addressing the needs of people with disabilities and their families.

ranging from, amongst others, developmental psychology (Geenen, 2007), psychoanalysis (Vanhee, 2007) and sociology (see Vranken chapter 6) in particular the sociology of childhood (see Morrow, 2001), to human rights law – including children's rights – and legal studies (see Vandenhole, 2008). Here, we discuss the issues of child poverty and children's rights from the perspective of social work.

The development of social work is one of change, but also one of continuity (Payne, 2005). In different European countries and since its origins, social work has been a class-specific discipline and practice focusing on people who are most likely to live at the bottom of the social hierarchy in Western society and who are characterised by a lack of resources, power and status (Craigh, 2003). On the one hand, social work focuses on social justice, striving for more social equality and a fairer redistribution of goods. The international definition of social work states that "the social work profession promotes social change, problem solving in human relationships and the empowerment and liberation of people to enhance well-being. Utilising theories of human behaviour and social systems, social work intervenes at the points where people interact with their environments. Principles of human rights and social justice are fundamental to social work" (Sewpaul & Jones, 2005: 218).

On the other hand, since the beginning of the 20th century, under cover of bourgeois philanthropy, the model of middle-class family life has been presented as the answer to social problems such as child poverty and criminality (Vanobbergen et al., 2006). Jones (2002) describes how the family has been seen as the principal institution that influences and informs the morality of children, and as the basic social unit for social work intervention. The reason for this is that explanations for child poverty were – and still are – particularly sought in poor and inadequate parenting, especially mothering (Garrett, 2007). In line with a logic of prevention, social work was developed as a welfare practice with overwhelming attention given to disciplining family life and to "good" parenting so as to protect children (Jones, 2002; Lister, 2003, 2006).

Moreover, after the post-war reform of welfare it became clear that policy changes following the Keynesian revolution throughout Europe did not result in a more equal material redistribution of society's resources, and that de facto problems of poverty and social inequality remained (Lawy & Biesta, 2006). Furthermore, Western welfare states have transformed themselves into workfare states over recent decades, in that they invest in "good" citizens who engage actively with the regular labour market (McDonald & Marston, 2005). In this welfare state model, social problems of long-term unemployment related to, amongst other factors, poverty, immigration, gender, illness and disability find a clear-cut solution in labour market activation (Bouverne-De Bie & van Ewijk, 2008).

Welfare reform is built around an emphasis on welfare-as-workfare strategies as the main route out of poverty (Wainwright, 1999).

The development from welfare to workfare holds the risk of casting a shadow over the social justice agenda of social work, and tends to nullify its ability to challenge structural social inequality (Powell, 2001), when it focuses on individual rather than structural change. This kite could be flying with regard to the notion of participation as well (Roose et al., 2009). On the one hand, it has been argued that a participatory approach can address issues of power inequality and result in an "almost unprecedented shift in power from manager and worker to client and a major culture change in the social care system" (Carr, 2007: 267). The assumption goes that participatory work results in a more qualitative kind of social work, especially in extremely difficult situations (Littell, 2001). On the other hand, central to social work's knowledge base and repertoire has been an individualistic approach towards families living in poverty. This individualistic perspective may continue under the participatory approach when the primary "causes" of clients' problems continue to be located solely in their individual characteristics and family relationships (Vanobbergen et al., 2006). The only scope for social work then seems to lie in governing, managing and policing child and family poverty as well as possible, and in not challenging society's social order fundamentally (Craigh, 2003). In light of this tendency, it is necessary to explore features of participatory social work with poor children and families. In what follows we deploy an analytical differentiation between the conceptualisation and interpretation of *participation as an instrument* and *participation as a point of departure*.

2. PARTICIPATION AS AN INSTRUMENT

In an individualistic and workfare approach to poverty, social policy makers take "good" citizenship as a starting point, which functions as a condition for individuals. This conditionality implies that citizens have no rights without responsibilities, and rights shift into social obligations to participate (Lorenz, 2001). Citizenship is conceptualised as an *achievement* (Lawy & Biesta, 2006), mainly based on distinctive norms and dichotomous categorisations that operate in every terrain of society. Social work in turn functions as an instrument of social policy. The emphasis of instrumental participation lies on the social integration and activation of poor people in order to prevent social problems (Roets, Roose & Bouverne-De Bie, 2009).

Participation is defined from a conception of citizenship where the ultimate goal for disadvantaged people in particular is to connect them individually with societal standards and interests (Roose & De Bie, 2007). "Bad" parents or the unemployed need to fit (and be fitted) into a prescribed outcome of becoming "good parents", being employable and employed. Participatory social work can acquire a moralising connotation as its focus lies on the active involvement of clients. Participation then becomes part of an instrumental strategy in which social work uses participation to realise its own prescribed goals (Roose et al., 2009). Recent research shows, for example, that parents who collaborate with social workers are seen as more responsible parents and their situation as less problematic (McConnell et al., 2006).

We argue that this instrumental approach to participation leads to a strange debate in social work practice: on the one hand the effects and outcomes of social work interventions have been defined beforehand, but on the other hand we get a debate on participation. This debate confines participation to the limits of what social workers think is important. Social work claims that participation is embedded in its practice but social workers decide what is necessary in the first place and clients become "accessories after the fact"; hence, participation may refer merely to a rhetorical change (Roose et al., 2009). In the end, it is the social worker who decides on what participation is, on who is allowed to or has to participate, and on who can or has to be empowered (Cruikshank, 1993; Baistow, 1994).

In an instrumental approach to participation, the power to define situations as a problem lies in the hands of social workers (Beresford, 2000). There is no shift in power, but rather a reinforcement of the power of the social worker, as the debate on the involvement of clients does not include their involvement in defining the problems at hand. This can be the case on an interpersonal level, an organisational level or on the level of social policy making. For instance, the social worker decides what the problem of the client is and then discusses with the clients how they can solve the problem together. In the context of child and family poverty, encounters with poor parents often result in a process of dialogue between social workers and parents concerning strategies to support them in the education of their children.

Although important, this process leaves out discussion of different possible perspectives on the situation and the way it is defined as a problem: it is the social worker who defines what the problem is (for instance, poverty as a consequence of a lack of parental skills) and clients are expected to be actively involved in solving it (for instance, the training of parents). The idea then is that clients have to learn to participate in social work so as to be empowered in a proper way, ostensibly on the basis that their voice is heard. Here, the language of participation operates as a camouflage technique: interventions are legitimised and clients become less resistant to the changes social workers deem necessary

(Postle & Beresford, 2007). For example, in Flanders, Belgium – as might be the case in other countries – we see an overrepresentation of poor children in child protection compared to the general population of children.

However, in the debate on child protection this overrepresentation is understood not as a problem of poverty but rather as the vulnerability of poor people to intrusive intervention. Hence we identify the need for participative social work as a means of making these interventions less intrusive. In the translation of the problem of poverty into a problem of lack of participation by poor people in the child protection system lies a risk of depoliticising poverty, by defining poverty into problems of the individual (Roose, 2006). This implies that parents are held responsible for the "good" education of their children, without social workers necessarily taking into account actual educational practices and the meanings derived from these practices within their broader societal context (Vandenbroeck, 2008). Practitioners might copy and produce views about "good parents" that enable them to handle social issues in a prescribed way:

> ...a 'good' parent raises children without public resources and ensures that children become self-sufficient or 'good' citizens who look after their own. They must achieve independence and 'break the cycle of poor parenting'. The fear that 'undeserving' parents will act as deficient role models transmitting poor parenting across generations and encouraging children's participation in a criminal underworld embeds the importance of 'breaking the cycle' of 'bad' parenting in practice (Dominelli et al., 2005: 1131).

Here we want to point to the underlying assumption that parents are to blame for the poverty and "bad" education of their children, devaluing them as future social assets, and to the risk of weakening solidarity with children and parents who live in poverty in the name of children's rights. In situations of child poverty, the rights of the child can easily serve to legitimise interventions targeted on those "bad" parents (Hamilton & Robert, 2000). The emotive power of children's rights easily makes a political imperative for intervention out of children's disadvantage and child poverty. Children's protection rights easily serve a sentimental view on child poverty based on a common morality concerning the protection of children as victims. Parents and children are then expected to participate in their own protection, but the notion of parental responsibility is used to legitimate the state's withdrawal from its responsibility to guarantee the rights of both parents and children (Fox Harding, 1996).

Policy makers and social workers who take an instrumental approach to participation in the name of children's rights might be concerned by the threat child poverty poses to the social order (Lister, 2003). The risk exists that clients who are perceived not to be participating as actively as social workers expect them to, or who are active in an undesirable way, are seen as non-participative, difficult, irresponsible and impossible in the social work process (Cowden & Singh, 2007). We follow Masschelein and Quaghebeur (2009), who argue that this creates new forms of exclusion. In the eyes of social workers, non-participation can easily be translated as problematic. This kind of dogmatic argumentation then ends the debate on structural strategies for combating child and family poverty. In a sense, participation is devoid of any meaning when problem constructions and outcomes of anti-poverty policy and social work have been defined in advance. In this interpretation, participation only enables poor people to toe the line set by a benevolent society. To challenge this instrumental argumentation, we need a different interpretation of participation.

3. PARTICIPATION AS A POINT OF DEPARTURE

In an alternative interpretation, *participation* is perceived *as a point of departure* in social policy making. In this approach the rational focus on the maintenance of social order shifts towards the basic principle of countering processes of structural marginalisation and on a redistribution of material resources and power (Doom, 2003). The emphasis lies on the question if and how social policy-making is democratic and related to the diversity of actual situations in which poor people live (Lister, 2002; Bouverne-De Bie et al., 2003). Processes of marginalisation and social exclusion are not merely about a lack of command over economic resources, but rather concern processes in society that lead poor parents and children to be excluded from a range of institutions, activities and environments (Redmond, 2008), such as education, employment, health care, and social and cultural initiatives (Wainwright, 1999). However, it is not only a question of being excluded from these institutions (and thus a question of equal opportunities, or of connecting people to the institutions), but also about being excluded from the debate on their meaning and the way interventions relate to a sense of human dignity (Bouverne-De Bie, 2003).

The rights of children and parents are then seen as being shaped through engagement in a participative process in which the definition and content of these rights are negotiated (Ife, 2005; Roose & De Bie, 2007). In this approach (see Roose & De Bie, 2007, 2008), the right of parents and children to participate functions as a starting point for dialogue and may take different and contextualised shapes. The actors involved may hold different perceptions, perspectives and meanings on the interpretation of rights and on what is in their best interest. As such, rights and responsibilities can be changed and renegotiated through (inter)actions in which contradiction and temporary consensus are vital elements. In this approach to participation the focus of social work shifts from the activation of clients to a reflexive praxis (Freire, 1972).

Central to this reflexive praxis is the question of what social work interventions mean for clients (and non-participative clients), rather than of their active involvement in the social work process (Roose & De Bie, 2007). For instance, what about a social work service that claims that it works in a very participative way but, when asked, answers that there are no clients who live in poverty because these people do not match their therapeutic approach? Is this participative social work? Maybe it is on a methodological level, concerning people who are accepted as valid clients. But the point is: does the social service have any idea what happens to the people they reject, or do they reflect on the significance of people in poverty who have no access to that service (Hubeau, 1995)? This implies, for instance, that the resistance of clients is not, by definition, considered as problematic but is possibly a meaningful sign: the only power clients have might be to disagree with the social worker (Pease, 2002; McLeod, 2007). It is important to see that reactions such as withdrawal, indifference, anger, fear, aggression, criminality and so on cannot simply be considered to be elements of deviant behaviour or problematic participation. These acts are meaningful and social workers have to look for their meaning. This implies a learning process for both social workers and clients that can develop in an interaction between people that is embedded in a set of relational questions, interests and concerns.

The question of the meaning of social work interventions refers not only to the meaning of the solutions social workers want to offer for the problems they encounter, but first and foremost to the construction process wherein situations are defined as social work problems. This means that participation cannot be enclosed in a prescribed method that makes social work more predictable. It is not about guidelines that tell you what to do to make the work easier and more effective. On the contrary, taking participation seriously makes social work more difficult and unpredictable, as it raises difficult questions that may turn the process of social work upside down as social workers have to take into account as relevant the different perspectives of the clients (Smith & Taylor, 2003), and have to learn how to share power in the identification and construction of problems and in stipulating joint action (Freire, 1972; Frankford, 1997).

While it is true that social workers and their agencies cannot overcome poverty directly, they *can* influence the way families experience poverty and related problems (Wainwright, 1999). They can influence the discourse on poverty and locate the problem of poverty in the individual or in society (Hawkins et al., 2001; Vojak, 2009). Social work and welfare initiatives that address the needs of poor service users must have at their core an understanding of poverty as a changeable social problem construction that can be deconstructed, shaped and reshaped in interaction and in dialogue with people who have direct experience of poverty (Lister, 2002). The fact that it might be hard to involve, for instance, poor children does not mean that their presence and their perspective should not be taken into account (Weinger, 2000; Walker, Crawford & Taylor, 2008). This does not mean that participation can be equated with doing what the client wants. It is about listening and looking at what is happening and about a continuous dialogue between social workers and clients, even if this results in a turbulent and discordant social work process.

4. CONCLUSION

Structural participation is fundamentally about questioning the meaning of social work interventions into the everyday lives of people, and to place this in the context of broader societal developments. The idea of participation and dialogue cannot be restricted within the boundaries of the dominant definitions of problems and solutions, such as the definition of poverty as an individual problem and individual empowerment as the solution. Otherwise, the social-political nature of problems risks disappearing from the discussion (Beresford, 2000).

Ife (2005) argues that social work must connect the macro and the micro, and needs to draw clear links between the private experiences of individuals and families and the socio-political context in which these experiences are located. The effort to understand a biography in all its uniqueness becomes the effort to interpret a social system (Booth, 1996). This applies both to theoretical understandings and also to ideas of action, and can be expressed as linking the personal and the political (Fook, 2002). Such an approach would move social work practitioners into an analysis of how individual and collective rights and responsibilities relate. It also implies a contextualised reading of children's rights (Roose & De Bie, 2007) and a questioning of the context in which children's rights – such as the right to participate – come to the fore. For instance, the sentimentalisation of the symbolic value of children is currently increasing (Pupavac, 2001). This process is believed to be related to the development of a risk society "in which the child is the source of the last remaining irrevocable, unchangeable primary relationship. Partners come and go. The child stays. Everything that is desired, but not realisable in the relationship, is directed to the child" (Beck, 1994: 118). This situation may lead to distrust of adults in general, but specifically of those who are seen as not fully capable of meeting the needs of children, such as parents in poverty.

A contextualised approach to children's rights has an eye for the possible negative consequences of some interpretations of children's rights, such as a diminishing understanding of the position of parents in poverty. A contextual approach to children's rights also demands that children's rights are linked to a human rights debate (Ife & Morley, 2002). Focusing exclusively on children's rights – and protection rights in particular – can easily serve a sentimental view of child poverty, based on a common morality for protecting children as victims of poor parents rather than on looking at children and adults in poverty as being subjected to processes of marginalisation and social exclusion. However, the human rights debate has to be rethought as well: human rights should be reclaimed and re-politicised (Gready & Ensor, 2005) so that they question and re-politicise the social problem construction of poverty rather than merely codify and smooth down power inequalities. Questioning the legitimacy of social, economic, cultural and political systems and social institutions, and the manner in which patterns of inequality constitute material circumstances and systematically disadvantage poor people – for instance through the demand that they must also be participative clients – should be fundamentally part of social work's agenda and of anti-poverty policy making (Bouverne-De Bie, 2003).

REFERENCES

BAISTOW, K. (1994), Liberation and regulation? Some paradoxes of empowerment, *Critical Social Policy*, (42): 3, 34–46.

BECK, U. (2007), *Risk Society. towards a new modernity*. London: Sage Publications.

BERESFORD, P. (2000), Service users' knowledges and social work theory: conflict or collaboration, *British Journal of Social Work*, (30): 4, 489–504.

BOOTH, T. (1996), Sounds of still voices: issues in the use of narrative methods with people who have learning difficulties, in: BARTON, L. (ed.), *Disability and society: emerging issues and insights*. London: Longman Sociology Series.

BOUVERNE-DE BIE, M., CLAEYS, A., DE COCK, A. et al. (2003), *Armoede en participatie*, Gent: Academia Press.

BOUVERNE-DE BIE, M. (2003), Een rechtenbenadering als referentiekader, in: BOUVERNE-DE BIE, M., CLAEYS, A., DE COCK, A. & VANHEE, J. (eds.), *Armoede en participatie*. Gent: Academia Press.

BOUVERNE-DE BIE, M. (2005), Armoede en kinderrechten, *Tijdschrift voor Jeugdrecht en Kinderrechten*, (3): 115–121.

BOUVERNE-DE BIE, M. & VAN EWIJK, H. (2008), *Sociaal werk in Vlaanderen en Nederland: een begrippenkader*. Mechelen: Wolters Kluwer Belgium.

CANTILLON, B. (2008), Armoede in België: over meten, weten, voelen en handelen, *De gids op maatschappelijk gebied*, (99): 3, 4–12.

CARR, S. (2007), Participation, power, conflict and change: theorizing dynamics of service user participation in the social care system of England and Wales, *Critical Social Policy*, (27): 2, 266–276.

CASHMORE, I. (2002), Promoting the participation of children and young people in care, *Child Abuse & Neglect*, (26): 837–847.
COWDEN, S. & SINGH, G. (2007), The 'user': friend, foe or fetish? A critical exploration of user involvement in health and social care, *Critical Social Policy*, (27): 1, 5–23.
CRAIGH, G. (2003), Poverty, social work and social justice, *British Journal of Social Work*, (32): 6, 669–682.
CRUIKSHANK, B. (1993), The will to empower: technologies of citizenship and the war on poverty, *Socialist Review*, (23): 4, 29–55.
DOMINELLI, L., STREGA, S., CALLAHAN, M. et al. (2005), Endangered children. Experiencing and surviving the state as failed parent and grandparent, *British Journal of Social Work*, (35): 7, 1123–1144.
DOOM, R. (2003), Essay van de open deur, in: BOUVERNE-DE BIE, M., CLAEYS, A., DE COCK, A. & VANHEE, J. (eds.), *Armoede en participatie*. Gent: Academia Press.
EU (EUROPEAN UNION) (2008), *European year for combating poverty and social exclusion (2010), strategic framework document: priorities and guidelines for 2010 European Year activities*. Brussels: European Commission Employment, Social Affairs and Equal Opportunities DG.
FOOK, J. (2002), *Social work. Critical theory and practice*. London/New Delhi: Sage Publications.
FOX HARDING, L.M. (1996), Recent developments in children's rights: liberation for whom? *Child and Family Social Work*, (1): 3, 141–150.
FRANKFORD, D.M. (1997), The normative constitution of professional power, *Journal of Health Politics, Policy and Law*, (22): 185–221.
FREIRE, P. (1972), *Pedagogy of the oppressed*. London: Continuum.
GARRETT, P.M. (2007). 'Sinbin' solutions: The 'pioneer' projects for 'problem families' and the forgetfulness of social policy research, *Critical Social Policy*, (27): 2, 203–230.
GEENEN, G. (2007), *Intergenerationele overdracht van gehechtheid bij Belgische moeders en kinderen die in extreme armoede leven: een meervoudige gevalsstudie*. Proefschrift in de Psychologie, Promotor J. CORVELEYN & K. VERSCHUEREN, Leuven: Katholieke Universiteit Leuven.
GREADY, P. & ENSOR, J. (2005), *Reinventing development? Translating rights-based approaches from theory into practice*. London/New York: Zed Books.
HAMILTON, C. & ROBERTS, M. (2000), State responsibility and parental responsibility: New Labour and the implementation of the United Nations Convention on the Rights of the Child in the United Kingdom, in: FOTTRELL, D. (ed.), *Revisiting children's rights. 10 years of the UN Convention on the Rights of the Child*. The Hague: Kluwer Law International, 127–147.
HAWKINS, L. FOOK, J. & RYAN, M. (2001), Social workers' use of the language of social justice. *British Journal of Social Work*, (31): 1, 1–13.
HOELSCHER, P. (2004), *A thematic study using transnational comparisons to analyse and identify what combination of policy responses are most successful in preventing and reducing high levels of child poverty*. Research report at University of Dortmund (Germany), commissioned by the DG Employment, Social Affairs and Equal Opportunities of the European Commission.

HUBEAU, B. (1995), De doorwerking van de sociale grondrechten in de Belgische grondwet: over de minimalisten en de maximalisten, in: VRANKEN, J., GELDOF, D. & VAN MENXEL, G. (eds.), *Armoede en sociale uitsluiting: jaarboek 1995*. Leuven: Acco.

IFE, J. (2005), Human rights and critical social work, in: HICK, S., FOOK, J. & POZZUTO, R. (eds.), *Social work: a critical turn*. Ontario: Thompson Educational Publishing.

IFE, J. & MORLEY, L. (2002), *Human rights and child poverty*. Fifth International Conference on the Child, Montreal, 23-25 May 2002, retrieved from: http://info.humanrights.curtin.edu.au/local/docs/HumanRights&ChildPoverty.pdf

JONES, C. (2002), Social work and society, in: ADAMS, R., DOMINELLI, L. & PAYNE, M. (eds.), *Social work: themes, issues and critical debates*. Hampshire/New York: Palgrave McMillan in association with The Open University.

LAWY, R. & BIESTA, G. (2006), Citizenship-as-practice: The educational implications of an inclusive and relational understanding of citizenship, *British Journal of Educational Studies*, (54): 34-50.

LISTER, R. (1997), *Citizenship: feminist perspectives*. London: MacMillan Press.

LISTER, R. (2001), New Labour: a study in ambiguity from a position of ambivalence, *Critical Social Policy*, (21): 4, 425-447.

LISTER, R. (2002), A Politics of recognition and respect: involving people with experience of poverty in decision making that affects their lives, *Social Policy & Society*, (1): 1, 37-46.

LISTER, R. (2003), Investing in the future citizens of the future: transformations in citizenship and the state under New Labour, *Social Policy & Administration*, (37): 5, 427-443.

LISTER, R. (2004), *Poverty: key concepts*. Cambridge: Polity Press.

LISTER, R. (2006), Children (but not women) first: New Labour, child welfare and gender, *Critical Social Policy*, (26): 2, 315-335.

LITTELL, J.H. (2001), Client participation and outcomes of intensive family preservation services, *Social Work Research*, (25): 103-113.

LORENZ, W., (2001), Social work responses to 'New Labour' in continental European countries, *British Journal of Social Work*, (31): 595-609.

MASSCHELEIN, J., QUAGHEBEUR, K. (2005), Participation for better or for worse, *Journal of Philosophy of Education*, (39): 1, 51-65.

McCONNELL, D., LLEWELLYN, G. & FERRONATO, L. (2006), Context-contingent decision making in child protection practice, *International Journal of Social Welfare*, (15): 230-239.

McDONALD, C. & MARSTON, G. (2005), Workfare as welfare: governing unemployment in the advanced liberal state, *Critical Social Policy*, (25): 3, 374-401.

McLEOD, A. (2007), Whose agenda? Issues of power and relationship when listening to looked-after young people, *Child and Family Social Work*, (12): 278-286.

MORROW, V. (2001), Using qualitative methods to elicit young people's perspectives on their environments: some ideas for community health initiatives, *Health Education Research: Theory and Practice*, (16): 3, 255-268.

PAYNE, M. (2005), *Modern Social Work Theory*. Houndmills/Basingstoke/Hampshire/New York: Palgrave McMillan.

PEASE, B. (2002), Rethinking empowerment: a postmodern reappraisal for emancipatory practice, *Sociale Interventie,* (11): 3, 29–37.
PLATT, L. (2005), *Discovering child poverty: the creation of a policy agenda from 1800 to the present.* Bristol: The Policy Press.
POSTLE, K. & BERESFORD, P. (2007), Capacity building and the reconception of political participation: a role for social care workers? *British Journal of Social Work,* (37): 143–148.
POWELL, F. (2001), *The politics of social work.* London: Sage Publications.
PUPAVAC, V. (2001), Misanthropy without borders: the international children's rights regime, *Disasters,* (25): 2, 95–112.
REDMOND, G. (2008). Child poverty and child rights: edging towards a definition, *Journal of Children and Poverty,* (14): 1, 63–82.
ROETS, G., ROOSE, R. & BOUVERNE-DE BIE, M. (2009), Ouders en kinderen in armoede en onderwijs: slechts kinderspel? Een verbindende visie op de samenwerking tussen ouder(s), onderwijs en welzijn, in: DE ZUIDPOORT VZW (ed.), *Ouder zijn: je bent en blijft het!* Verslagboek studiedag 27 maart 2009. Gent: vzw De Zuidpoort.
ROOSE, R. (2005), Participatieve hulpverlening: bron of fata morgana? in: VAN BUYTEN, K. (red.), *Participatierechten van kinderen. Verzamelde commentaren.* Cahier 25. Gent: Academia Press, 219–240.
ROOSE, R. (2006), *De bijzondere jeugdzorg als opvoeder.* Gent: Academia Press.
ROOSE, R. & DE BIE, M. (2007), Do children have rights or do their rights have to be realised? The United Nations Convention on the Rights of the Child as a frame of reference for pedagogical action, *Journal of Philosophy of Education,* (41): 3, 431–443.
ROOSE, R. & DE BIE, M. (2008), Children's rights: a challenge for social work, *International Social Work,* (51): 37–46.
ROOSE, R., MOTTART, A., DEJONCKHEERE, N. et al. (2009), Participatory social work and report writing, *Child & Family Social Work,* (14): 3, 322–330.
SCHON, D.A. (1983), *The reflective practitioner.* London: Temple Smith.
SEWPAUL, V. & JONES, D. (2005), Global standards for the education and training of the social work profession, *International Journal of Social Welfare,* (14): 3, 218–230.
SMITH, A. & TAYLOR, N. (2003), Rethinking children's involvement in decision-making after parental separation, *Childhood,* (10): 201–216.
STEENSSENS, K., AGUILAR, L.M., DEMEYER, B. et al. (2008), *Kinderen in armoede. Status quaestionis van het wetenschappelijk onderzoek in België.* Leuven: Interuniversitaire Groep Onderzoek en Armoede (IGOA) vzw.
UNICEF (2000), *Child poverty in rich nations.* Florence: United Nations Children's Fund, Innocenti Research Centre.
UNICEF (2007), *Child poverty in rich countries.* Florence: United Nations Children's Fund, Innocenti Research Centre.
VANDENBROECK, M. (2008), Early childhood education and poverty, in: SCHRONEN, D. & URBÉ, R. (eds.), *Socialalmanach 2008. Schwerpunkt Kinderarmut und Bildung.* Luxembourg: Caritas Service de Recherche et Développement, 255–267.

VANDENHOLE, W. (2008), Conflicting economic and social rights: the proportionality plus test, in: BREMS, E. (ed.), *Conflicts between fundamental rights*. Antwerp: Intersentia, 559-589.

VANHEE, L. (2007), *Weerbaar en broos: mensen in armoede over ouderschap. Een verkennende kwalitatieve studie in psychologisch perspectief*. Proefschrift in de Psychologie, Promotor J. CORVELEYN. Leuven: Katholieke Universiteit Leuven.

VANOBBERGEN, B., VANDENBROECK, M., ROOSE, R. et al. (2006), 'We are one big, happy family': beyond negotiation and compulsory happiness, *Educational Theory*, (56): 4, 423-437.

VOJAK, C. (2009), Choosing language: social service framing and social justice, *British Journal of Social Work*, (39): 936-949.

VRANKEN, J., DE BOYSER, K. & DIERCKX, D. (2004), *Armoede en sociale uitsluiting, jaarboek 2004*. Leuven: Acco.

VRANKEN, J. (2004), Algemene inleiding: jaarboek (13 jaar) kijkt terug op verslag (10 jaar), in: VRANKEN, J., DE BOYSER, K. & DIERCKX, D. (eds.), *Armoede en sociale uitsluiting, jaarboek 2004*. Leuven: Acco.

VRANKEN, J. (2008), *Armoede en sociale uitsluiting, jaarboek 2008*. Leuven: Acco.

WAINWRIGHT, S. (1999), Anti-Poverty strategies: work with children and families, *British Journal of Social Work*, (29): 477-483.

WALKER, C. & WALKER, A. (2002), Social policy and social work, in: ADAMS, R., DOMINELLI, L. & PAYNE, M. (eds.), *Social work: themes, issues and critical debates*. Hampshire/New York: Palgrave McMillan in association with The Open University.

WALKER, J., CRAWFORD, K. & TAYLOR, F. (2008), Listening to children: gaining a perspective of the experiences of poverty and social exclusion from children and young people of single-parent families, *Health and Social Care in the Community*, (16): 4, 429-436.

WEINGER, S. (2000), Economic status: middle class and poor children's views, *Children & Society*, (14): 2, 135-146.

5. CHILDREN IN PUBLIC CARE IN ENGLAND: WELL-BEING, POVERTY AND RIGHTS

Nina Biehal and Gwyther Rees

INTRODUCTION

This chapter considers the well-being, poverty and rights of children and young people who are "looked after" in public care in England, and situates the discussion of each of these issues in the context of the wider research on these topics.[1] This group typically live in family foster care, but around one in ten of them live in residential children's homes. Due to their very troubled backgrounds before entering care, initial well-being may be low for many of these children. Their past adversities may also have repercussions for their ongoing well-being. Although substitute care may provide some compensation for past adversities, there are continuing concerns that public care may also compound these difficulties instead of compensating for them. Many of these children have experienced poverty within their families before entering care and there is concern that they are at particularly high risk of poverty after leaving care, in late adolescence. They are therefore highly vulnerable to a range of adversities prior to, during and after the time they are in care.

Our focus on this particularly vulnerable group raises key dilemmas in relation to children's well-being and children's rights which may be of relevance to other marginalised groups, such as disabled or asylum-seeking children, as well as to other children and young people in the wider community. How far societies ensure the well-being of the most marginalised groups of children may be an important indicator of how seriously they take their responsibilities towards all children.

---

[1] There are some differences in policy and in the provision of child welfare services between the four jurisdictions within the UK (England, Wales, Scotland and Northern Ireland). For the most part, this chapter focuses on England.

1. CHILD WELL-BEING

The issue of young people's well-being is increasingly being debated both in the UK and internationally. Recent comparative research has suggested that there are substantial variations in well-being between nations and that young people in the UK fare relatively poorly in comparison with their peers in most other European countries (Bradshaw & Richardson, 2009). This has led to considerable concern about the well-being of children and young people in the UK.

At a policy level, in 2003 the UK Government initiated the *Every Child Matters* policy agenda to improve the well-being of children and young people in England. *Every Child Matters* serves both as a statement of national policy goals in relation to all children and as a guide to local policy and service development. Developed in the context of the government's performance indicators for local authorities, *Every Child Matters* also sets out a framework against which child well-being may be assessed, both in relation to all children and, in some instances, in relation to specific groups such as children in low-income households or in public care. The *Every Child Matters* outcomes framework entails a conceptualisation of well-being:

– Being healthy
– Staying safe
– Enjoying and achieving
– Making a positive contribution
– Achieving economic well-being

This framework reflects some of the complexities and tensions in thinking about the concept of well-being in relation to children and young people in particular. First, there is a tension between thinking about young people's current well-being and about their future well-being as older young people and adults (which has been termed "well-becoming"). Second, there is a tension between young people's own well-being and how they contribute to the well-being of others and of society as a whole.

The above tensions, in the UK at least, are reflected in some of the current public debates about young people. First, there is considerable concern about young people's educational achievement and health – and the potential implications that these may have for their future well-being. Second, young people are often portrayed in the media as troubled (e.g. rising mental health problems) and/or troublesome (e.g. involvement in anti-social behaviour).

5. Children in public care in England: well-being, poverty and rights

Historically, these same tendencies have to some extent been a feature of social surveys of young people in the UK and elsewhere. There has been a substantial amount of work on negative aspects of well-being such as mental ill-health. However, as has often been noted in the literature, the concept of well-being encapsulates more than just the "absence of negative factors, as is true of most measures of mental health" (Diener, 1984: 543–544). There is also a considerable body of research on behavioural issues, which may have an impact on the well-being of young people and society (e.g. drug use and offending). In comparison, there has been relatively little focus on young people's positive well-being in the present, and in particular on subjective perspectives on well-being.

Over the past decade, internationally, this situation has begun to be addressed by research (see Ben-Arieh, 2008 for an historical overview) which has attempted to shift the focus of well-being research in three ways:

- From negative indicators to positive indicators
- From proxy behavioural indicators to subjective measures
- From the future to the present.

These three trends have begun to redress the balance and focus more strongly on subjective well-being, measured positively.

This is not to suggest that objective indicators, negative well-being, and issues of behaviour and well-becoming are not important components of the overall concept. However, the debate has lacked sufficient balance in acknowledging young people's feelings about how they experience their lives in the present as a valid and important aspect of well-being. Moreover, concepts of young people's well-being have also been developed from concepts which apply to adults without consideration of the potential differences according to age (Fattore et al., 2009).

As part of this trend towards a more child-centred approach to well-being, the Children's Society and the University of York have developed a research programme which attempts to build a concept of well-being based on young people's perspectives as well as on the available academic work, and then to measure well-being through a self-report survey of young people.

The first qualitative phase of this research project, which gathered views from 8,000 young people (see The Children's Society, 2006), highlighted some important issues, from young people's viewpoints, that have been relatively underplayed in some previous frameworks of child well-being. Thus, whilst young people recognised the important contributions which issues such as education, health and their own behaviours make to their current and future well-being, their most fundamental concerns were with the quality of their relationships with family and friends. In addition, they emphasised a set of crosscutting conditions, including fairness, respect and freedom (balanced with safety), which they felt were important aspects of their relationships and environments.

Building on this first phase, and also taking into account international well-being literature, a draft conceptual framework was developed. This framework then formed the basis of the first wave of a survey of a representative sample of 7,000 young people aged 10 to 15 in 2008. The first findings from the research were published in early 2010 and confirmed the primary importance of family relationships and concepts of autonomy and freedom as key components of young people's well-being (Rees et al., 2010). The intention is to repeat this survey every two years and also to undertake additional survey and research work with more marginalised groups of young people.

1.1. THE WELL-BEING OF CHILDREN IN PUBLIC CARE

Since the mid-1980s, research on young people leaving foster and residential care (between the ages of 16 and 18 years) has shown that they fare extremely badly in comparison with young people in the wider population. Education and employment outcomes are poor, increasing the risk of poverty and homelessness. Care leavers often lack basic life skills and informal social support, and are more likely to become parents at an early age (Biehal et al., 1995; Stein & Carey, 1986). Other research has suggested that they are also over-represented among the prison population (Prison Reform Trust, 1991).

These concerns about the well-becoming of young people in public care have been compounded by growing concern about their well-being during the time they are in their care placements. From the mid-1990s, it has become increasingly clear that the prevalence of mental health problems is much greater among the care population than among children and young people in the wider community (McCann et al., 1996). Studies have consistently identified high rates of conduct disorder among children in public care (Hill, 2009). A UK study of a representative sample of 1,039 children in public care reported that 45 percent had a mental disorder, a prevalence rate four to five times higher than for the wider population of children and young people under the age of 18 (Meltzer et al., 2000; Meltzer et al., 2003). This national survey found that, *compared to children living in disadvantaged families*, they were nine to ten times more likely to have a conduct disorder, two to three times more likely to be depressed, three to four times more likely to have a hyperkinetic disorder and 11 to 12 times more likely to have a post-traumatic stress disorder (Ford et al., 2007).

Research has also shown that emotional and behavioural problems among children in public care are often associated with their experience of movement between foster and residential placements. One study of young people leaving care found that 40 percent had moved placements four or more times and ten per cent had moved more than ten times (Biehal et al., 1995). Mental health difficulties, particularly conduct and hyperkinetic disorders, may increase the risk of placement breakdown, but placement instability may be a cause as well as a consequence of children's difficulties. Placement stability has been found to be a predictor of well-being for children in foster care, although the direction of causation is difficult to establish (Biehal et al., 2010; Rubin et al., 2009).

The low educational achievement of children in public care may have consequences for both their well-being and their well-becoming. Inadequate corporate parenting involving low expectations, a failure to prioritise education and the disrupted schooling that may result from placement instability all contribute to this picture of low attainment (Berridge & Saunders, 2009). There is also evidence that compared to other young people, they are twice as likely to be drawn into the criminal justice system.

The experience of multiple adversities prior to entry to care may therefore be compounded by experiences while *in* care, such as placement instability or educational failure. Together, pre-care and in-care risk factors may have an impact on the quality of young people's well-being during the time they are in care, unless they are supported to settle in placements and receive the mental health or educational support they might need. These factors may also have consequences for their well-becoming, as international research has shown that young people who have been in care are at high risk of social exclusion in early adulthood and beyond (Stein, 2009). These concerns have led not only to the development of new legislation and services to prepare and support young people *leaving* care, but also to increasing attention to the support to improve their well-being during the time they are *in* foster and residential care. Thus, concerns about the well-becoming of young people has strengthened attempts to improve their well-being during the time they spend living in public care.

2. CHILD POVERTY AND WELL-BEING

In a study exploring the relationship between child poverty and well-being in 25 European countries, Bradshaw and colleagues examined a number of different approaches to measuring child poverty, including relative income poverty, joblessness, subjective poverty and deprivation (as well as combinations of two or more of these) (Bradshaw et al., 2007). This study used a measure of well-being which comprised eight domains: material situation, housing, health, education, children's relationships, risk and safety, civic participation and subjective well-being. They found a link between poverty and well-being, but discovered that this relationship between the two operated mainly through the dimension of deprivation, which they defined as "the lack of a necessity or object required to maintain an acceptable standard of living". Another cross-national study of 23 rich countries (in Europe, North America, Australasia and the Far East) found that child well-being was lower in countries with a greater degree of income inequality and a higher percentage of children in relative poverty. That study concluded that improvements in child well-being are likely to depend more on reducing inequality than on increasing economic growth (Pickett & Wilkinson, 2007).

In some studies the concepts of poverty and well-being may be conceptually distinct, while in others they may overlap. This depends on how well-being is defined, as it may be measured in terms of social indicators, self-report or a combination of the two. One the one hand, multi-dimensional measures of well-being which include a dimension referring to material deprivation may be viewed as broader than the concept of poverty, but not entirely distinct from it. Such a conceptualisation of well-being would be consistent with Gough and McGregor's contention that the obverse of well-being is not simply poverty but a wider conceptualisation of ill-being (Gough & Mc Gregor, 2007). On the other hand, measures of well-being which do not include a domain referring to material circumstances may be viewed as conceptually distinct from poverty. Indeed, it would be more meaningful to test for associations between poverty and a measure of well-being that was, in this way, conceptually distinct from poverty.

2.1. WELL-BEING AND POVERTY FOR CHILDREN IN PUBLIC CARE

It has long been known that children and young people in public care often come from families living in poverty (Bebbington & Miles, 1989). The harm associated with poverty is known to include increased risk of detected maltreatment, teenage pregnancy, disorganised parenting and substance misuse, all of which increase the likelihood that children might enter public care. The link between poverty and entry to care is complex, however. A recent study of nearly 100,000 children admitted to care in Denmark over a period of 23 years has shown that the link between socio-economic status and the risk of entry to care is complex and varies by age (Andersen & Fallesen, 2010). Poverty is only one of many adversities experienced by children who eventually enter public care, but it may interact with other factors to increase the risk of entry to care. In England, a study of a cohort of 14,256 children found that among children under the age of six, indicators of poverty significantly increased the likelihood that a child's name would be entered on the child protection register, which in itself would increase the likelihood of entry to care.

However, the odds that children from families living in poverty would enter care were reduced once parental background factors were taken into account. Four key parental factors increased the risk of maltreatment: age (that is, early parenthood), low educational achievement, a history of mental health problems and a parental history of abuse. The researchers suggested that these parental factors were mediated through the socio-economic environment of these families. They also argued that the association between poverty and child maltreatment might be influenced by referral bias, as thresholds for referral to social services might be lower for young and poor parents (Sidebotham and Heron, 2006). The link between poverty and entry into care therefore appears to be indirect, at least in England.

The threshold for entry into the English care system is higher than in many other countries, with care provided mainly for children who have experienced multiple and severe adversities, in most cases involving serious abuse or neglect (Thoburn, 2009). Many of their parents have apparently intractable substance misuse or mental health problems, and many have experienced domestic violence in their families. Poverty may be a risk factor for entry into care, but it generally operates indirectly. The high threshold for entry into the English care system is partly explained by concerns about the high costs of looking after children in foster or residential care, particularly in the context of the trend, over the last 15 years, for children to enter care younger and remain longer (Department of Health, 2001; Rowlands & Statham, 2009). A more positive reason for the residual nature of the care system in England is the focus on the provision of family support, which has been a statutory duty since 1963 (as discussed below in the section on English child care policy). Where poverty is the principal cause of children's difficulties, in the absence of other serious adversities children's services aim to provide support to keep children within their families rather than take them into care.

Once children have entered care, what is the link between well-being, well-becoming and poverty? Although there have been no recent surveys of this, research on children who enter care suggests that many may have experienced deprivation prior to entry. However, they cease to experience this deprivation directly once they move into residential or foster homes. Nevertheless, the experience of deprivation prior to entry into care may have continuing effects for some children. The strength of these effects is likely to depend to some extent on the age at which children enter care and how long they stay. For example, a study of 16–18 year olds leaving care found that those who entered care during adolescence did worse on a range of measures, including education, training, employment and housing, than those who entered when younger (Dixon et al., 2006).

For young people who enter care during adolescence, it may be harder to deal with attention problems, to break well-established patterns of disengagement from education, build confidence and encourage motivation and so provide some compensation for earlier educational problems. Where this cannot be achieved, young people are at greater risk of unemployment, and hence of poverty, after leaving care. Similarly, the mental health problems that children bring with them into public care may not only cause distress in the present but may also increase the risk of poor mental health and associated social isolation, joblessness and poverty in the future.

Research has shown that young people leaving care at the age of 16 and over are likely to experience multiple disadvantages, including poverty, as a result of both their poor family backgrounds and damaging experiences within their families and of the failure of care to provide stability and compensate for these adversities. For some, the risk of poverty is compounded by low educational attainment and the significantly higher likelihood, compared to other young people, that they will become parents during adolescence (Biehal et al., 1995; Stein, 2005).

Bradshaw and colleagues found no relationship between educational attainment and overall well-being, measured along the seven domains outlined above (Bradshaw et al., 2007). They argue that educational attainment should be viewed as a measure of well-becoming rather than well-being. However, for a vulnerable group such as children in care, for whom emotional, behavioural and educational difficulties may often be a consequence of the severe adversities they have experienced, achievement at school may help to build confidence, self-esteem and an overall sense of subjective well-being.

The well-being that may result from improvements in these more objective, measurable indicators of well-being, in the domains of education or mental health, may have both hedonic and eudaimonic aspects. The hedonic aspect of well-being is often referred to as subjective well-being, which may include perceptions of happiness, life satisfaction and positive and negative affect. The eudaimonic aspect of well-being concerns psychological well-being, which may comprise a sense of purpose or meaning and personal growth (Rees et al., 2010). Children and young people may feel both happier and less anxious if they receive help with mental health, learning or peer problems. Equally, tackling these difficulties may increase their sense of purpose, achievement, self-esteem, personal growth and engagement with others, an aspect of well-being which is broadly captured by the "enjoy and achieve" dimension of the *Every Child Matters* framework.

Until relatively recently, however, social workers sometimes tended to focus more on children's well-being than on their well-becoming, paying relatively little attention to ensuring that children benefited fully from universal services such as health and education. Emotional and behavioural difficulties or a failure to engage in education were viewed as problems that were only to be expected, given the children's past adversities. Furthermore, in relation to education there was often a well-intentioned reluctance to seek additional support for children, as it was sometimes felt that they were too troubled to have the emotional resources to focus on learning, an attitude which fuelled a general culture of low expectations (Wade et al., 1998; Jackson, 1998; Berridge, 2007; Berridge & Saunders, 2009). This failure to support children's education not only reinforced patterns of failure, harming their subjective well-being while in care, but contributed to their difficulties in obtaining employment as young adults, thus increasing the risk of future poverty.

Although over the last twenty years or so sociologists of childhood have suggested that greater attention should be given to well-being, attention to well-becoming remains important, at least for children in public care. A focus on both well-being and well-becoming may generate a positive chain of events that may be protective against future poverty. If children in care are to avoid poverty and other poor outcomes in early adulthood, it is essential to focus both on their well-being now *and* on their well-becoming in the longer term.

3. THE RIGHTS OF CHILDREN IN AND ON THE EDGE OF PUBLIC CARE

There have been long-standing debates about the balance of rights and responsibilities between parents and the state. For maltreated children, including those who enter public care, the balance between the two at any point in time may also highlight potential tensions between the rights of children and the rights of their parents. We first assess the situation of children's rights in England and next examine the question of parental responsibility and the role of the state and the implications of these for the rights of vulnerable children. We then consider the particular relevance of the rights to protection, non-discrimination and participation, as set out by the UNCRC, to children in public care.

3.1. CHILDREN'S RIGHTS IN ENGLAND

The United Nations Convention on the Rights of the Child (UNCRC) upholds children's rights to life, survival and development, setting out their rights to protection, provision and participation. It might be argued that the fulfilment of these rights is effectively the same as meeting children's needs for them. Further, meeting children's needs is likely to improve their well-being. Although "rights" are not reducible to "needs", the two concepts are closely related. Gough and Doyal have argued that certain human needs are universal. In their formulation, universal human needs are conceptualised along two dimensions: survival/health and autonomy/learning (Doyal, 1991). The core rights to life, survival, protection and provision might be viewed as addressing children's needs for survival/health, while the rights to development and participation would help to meet their needs for autonomy/learning. Children's rights to protection, provision and participation are therefore firmly grounded in the basic human needs for life, growth and development (Reading et al., 2009).

The UK has responded to the UNCRC by taking a sectoral approach to law reform, gradually examining legislation in different areas (Innocenti Research Centre, 2007). As a result, attempts to ensure that children's rights in the UK are consistent with the Convention have been somewhat piecemeal. The Children's Plan for England, published by the English government in 2007, refers to the UNCRC but is not framed by it. On the whole, the government's attempts to ensure children's well-being and children's rights have focused on specific areas, such as child protection and participation in education. The UK government currently has no plans to incorporate the UNCRC into UK law (Children's Rights Alliance for England, 2009). A consequence of this is that a number of the rights set out in the UNCRC are not currently enforceable in the UK.

At present, there are a number of concerns about the treatment of children and young people in England. These include concerns about the demonisation of young people, the operation of the youth justice system and the treatment of asylum-seeking children. The generally negative representations of young people in sections of the media, which frequently portray them as "feral" or menacing, has some resonance with government policy on anti-social behaviour. While anti-social behaviour by both children and adults is a genuine problem, and children themselves are often the victims of anti-social behaviour, there are concerns that the nature of the government response has reinforced representations of young people as a threat. The government has also been criticised by the UN Committee on the Rights of the Child and the Human Rights Commissioner of the Council of Europe for the low age of criminal responsibility (ten years) and for its use of custody for young people, as England and Wales lock up more young people than most other European countries. In addition, there are serious concerns about the detention of refugee and asylum-seeking children.

3.2. ENGLISH CHILD CARE POLICY: BALANCING THE RIGHTS OF CHILDREN AND PARENTS

For at least sixty years, child welfare policy in England has oscillated between a focus on the protection of children and the rights of parents. Since the passing of the Children Act 1948 there has been a continuing tension between state paternalism, prioritising the safety of children, and the defence of the birth family. In some periods policy has favoured a legalistic, protectionist approach to children while in others there has been a stronger focus on providing support to families with a view to preventing children's entry into public care whenever possible (Packman & Jordan, 1991; Packman, 1993). This has been characterised as an ongoing conflict between protagonists of the "state as parent" and the "kinship defenders", with the latter arguing for greater support to families in order to prevent admission into public care (Fox Harding, 1991).

From the late 1940s, there was increasing recognition of the link between poverty and child maltreatment and a questioning of the wisdom of removing maltreated children from home rather than providing support to their families to prevent their removal. The increasing numbers of children entering care during the 1950s and the rising costs associated with this increase, together with growing recognition of the psychological consequences of separation for children, contributed to the new focus on prevention evident in the Children and Young Persons Act 1963. This was the first legislation to set out a statutory duty to provide assistance to families in order to diminish the need to receive children into care. The concept of prevention embodied in the 1963 Act was quite broad, encompassing not only the prevention of entry into care but also the prevention of cruelty and neglect.

During the 1970s the pendulum began to swing the other way. Concerns were expressed about the psychological consequences for children of "drifting" in care, or repeatedly moving into and out of care, and a greater emphasis was placed on their need for permanent placement to ensure stable relationships (Fanshel & Shinn, 1978; Maluccio & Fein, 1983; Goldstein et al., 1973; Rowe & Lambert, 1973). During this period, influential commentators promoted the idea that state intervention in family life should be minimal but that, once parenting was found to be severely impaired, decisive action should be taken by the state. This action might include the severance of ties with the biological family and encouragement of the development of new psychological ties to substitute parents, including adoptive parents (Goldstein et al., 1973). Perhaps even more influential in the UK, though, was the series of public enquiries from 1974 onwards into child deaths resulting from physical abuse (Parton, 1991; Secretary of State for Social Services, 1974). The new focus on children's needs for protection, stability and permanence brought with it a return to an (implicit) prioritisation of children's rights over the rights of parents.

By the 1980s, however, the balance between a focus on protecting children and a focus on parents' rights in child care policy had begun to shift yet again. A number of studies argued for greater attention to supporting families in order to prevent family breakdown and a review of child care law questioned the distinction between the needs of the child and the needs of the family (Department of Health and Social Security, 1985; Parker, 1980; Fisher et al., 1986; Holman, 1988). During the same period, a collection of pressure groups that constituted the parents' rights lobby, such as the Family Rights Group, campaigned to re-establish a commitment to preventive work.

These developments helped to shape the Children Act 1989, which placed a duty on local authorities to safeguard and promote the welfare of "children in need" and to promote their upbringing by their families. Parents of "children in need" were encouraged to look to local authorities for non-stigmatising support services, without fear of a loss of parental responsibility. Indeed, the continuity of parental responsibility was emphasised by the Children Act 1989, thus satisfying both conservative and liberal commentators. The focus on supporting families and the promotion of parental responsibility reflected the Conservative anti-collectivist policies of the time, which aimed to reduce the role of the welfare state and reassert the role of the family in taking responsibility for its members. In line with this philosophy, the emphasis was on reducing state intervention in the private sphere of family life. At the same time, this policy satisfied those with more liberal views regarding the need for support to disadvantaged families to enable them to care adequately for their children. Thus for different reasons, policy on family support appealed to "kinship defenders" across a wide political spectrum (Biehal, 2005).

The Act sought to integrate the duty to protect children from harm and ensure their proper development with the duty to support families. It represented an attempt to balance the rights of children with the rights of parents, but nevertheless prioritized the rights of children through its core principle that the welfare of children should be the paramount consideration. Although the notion of children's rights is not explicit in the Children Act 1989, the "paramountcy principle" at its core nevertheless privileges their welfare. The Act does not refer to parents' "rights" but instead to their responsibilities to their children. If children enter public care with the voluntary agreement of parents, they retain full parental responsibility. When children are looked after under a legal order, parental responsibility is shared between parents and the local authority caring for the child.

As these developments in English child care policy indicate, there has long been a fine balance between ensuring the rights of children to safety and stability and respecting the rights of parents. The provisions of the Children Act 1989 are consistent with the core rights specified by the Convention to protection, provision and participation and also with its general principles that the best interests of the child should be a primary consideration and that the views of the child should be respected. Subsequent child care legislation in England has built on or amended the Children Act 1989 but has not deviated from its principles. Successive acts of parliament and public policy on children since 1989 have given increasing prominence to the requirements of the UNCRC, such as giving children a voice in assessment and in court proceedings (Reading et al., 2009).

The UNCRC also specifies the government's duty to support parents in bringing up their children. Recently, rights-based approaches have been used in the courts to champion the rights of parents over those of children. There is emerging evidence that barristers sometimes advise birth parents to use the Human Rights Act 1998 to make legal challenges to local authorities' plans for adoption (Lowe et al., 2002; Ward et al., 2006; Munro & Ward, 2008; Biehal et al., 2010). Thus parents' claims to the right to family life may override the UNCRC principle that the best interests of children should be the main consideration, with the result that children who can never safely return home may lose their chance to the greater security provided by adoption, instead remaining in public care. Nevertheless, although the relationship between parents' and children's needs and rights may be complex for children who are maltreated and/or separated from their families, it is important to acknowledge that in many cases their needs and wishes may be compatible.

3.3. PROTECTION

It has long been accepted that definitions of abuse and neglect are culturally determined, and this recognition has underpinned relativist discourses on maltreatment. Such relativist approaches may also be informed by continuing uncertainty as to whether abuse should be defined in terms of acts or outcomes. Definitions based on acts are usually legal definitions – for example, a sexual act with a young child would be defined as illegal – but in other circumstances an act may not be perceived as harmful by the child, or no discernable harm may be caused (Finkelhor, 1994; Forrester & Harwin, 2000). Furthermore, some libertarian commentators on children's rights have challenged the view of "child-savers", who consider maltreatment to be an absolute harm from which children should be protected by adults. Liberationists (e.g. Holt, 1975; Franklin, 1989) have viewed concern with the protection of children as a form of paternalism, which restricts their autonomy, prioritising their right to autonomy over their right to protection (Fox Harding, 1996).

Such analyses appear to ignore the power relations within families in which children experience severe maltreatment, which may deny children any real autonomy at all. Similarly, whether a child views the harm caused by abuse or neglect as negligible may be irrelevant, as some maltreated children may come to accept emotional or physical harm or neglect as "normal", never having known anything else. Furthermore, their interpretation of their own situations may be heavily influenced by parental narratives and by their own complex feelings towards their parents. This is not to argue that children's views are themselves irrelevant. Listening to children is essential. However, assumptions about the rational and coherent nature of expressed opinions ignore the ambivalence, fear and confusion that may potentially underlie children's (and indeed adults') expressed opinions.

In relation to maltreatment, the Convention tackles this tension between absolute and relativist discourses of needs and rights. As well as the right to life (article 6), the UNCRC grants children protection from all physical, sexual and mental abuse, neglect and exploitation (article 19). It thus upholds their absolute right to protection. Like adults, children have basic human needs for survival/health, including an absolute need for protection from harm, and this protection is granted to children by the UNCRC as a right. Reading and colleagues have argued that a children's rights approach within different social and cultural contexts may be an important strategy for resolving the tension between absolute and relativist responses to maltreatment (Reading et al., 2009). This would certainly be true in relation to the approach taken by the UNCRC, but not in relation to liberationist approaches which prioritise autonomy over protection.

3.4. NON-DISCRIMINATION

A children's rights approach may also be a helpful framework within which to consider the problems that children in care experience in relation to education and criminalisation, as outlined above in our discussion of well-being. The high rates of exclusion from school for children in care and the evidence of long delays in the provision of schooling for these children run counter to article 28 of the UNCRC, which requires governments to make education accessible to all children. Whereas local authorities may take parents to court if they fail to ensure that their children attend school, for children in care it is sometimes the local authority itself which fails to provide a school place when children are in public care (Wade et al., 1998). This failure to provide education for all children in care may be viewed as a form of discrimination, as may a failure to provide appropriate and sufficient support to improve their attainment. Taking a life course perspective, it is clear that educational failure may increase the risk of later unemployment, poverty and homelessness, as argued above. The English government has acknowledged that many children in care are let down by the system that is supposed to look after them and has introduced a number of initiatives in response to these problems.

In relation to the disproportionate number of children in care involved with the criminal justice system, there is some evidence that, to some extent, discrimination may play a part. It has been suggested that young people in care are disproportionately criminalised, as they are sometimes referred to the police for behaviour which, in other circumstances, might be dealt with informally by a parent (Darker et al., 2008). Some therefore enter the formal criminal justice system as a result of minor offences in circumstances which would not lead to this outcome for children living with their families. They may also find themselves bullied by their peers simply because they are in care. Such forms of discrimination may arise specifically as a consequence of children's care status (Lindsay, 1998).

3.5. PARTICIPATION IN DECISION-MAKING

More positively, the English law in relation to children in public care is consistent with the requirement of article 12 of the UNCRC that children's views should be given due weight in decisions that affect them. The Children Act 1989 requires that children's "wishes and feelings" must be considered when making any decision about them. Children may also apply for contact with a parent, or apply for contact to be refused, and must have their complaints considered. However, the legislation frames these as "requirements" for the local authority, not as rights of the child.

The critical issues here are, first, whether such "requirements" are enforceable by children and second, whether they should be. The perspectives of liberationist writers on children's rights are of relevance to both of these issues. While liberationists have been criticised for their failure to recognise that some children do need special protection, they have made a valuable contribution in arguing that children themselves should be empowered in order to ensure that adults fulfil their responsibilities to them.

The issue of *the extent* to which children should be able to influence decisions about their own lives is more complex. To begin with, it raises the much-discussed question of competence, on which liberationists, some feminist writers and others have varying views (Alderson, 1992; Lindsay, 1992). More problematic, though, is the fact that the rationalist discourse of children's rights positions children as independent beings, disembedded from relationships with their families and carers, who make rational choices about the nature of their care. For adults and children alike, both the focus on well-being and the focus on positive rights that can be claimed by individuals have emerged in a contemporary context which emphasises self-responsibility, autonomy and freedom (Sointu, 2005; Rose, 1999). Such conceptions of self-responsibility imply the existence of a rational, unified self disconnected from social relationships and, importantly, from relations of power (Biehal & Sainsbury, 1991). Within this framework (inter)dependency, for both adults and children, may be viewed negatively, rather than accepted as the reality of people's lives.

For children separated from their parents following traumatic experiences, emotions about their birth families may be complex and may influence both their relationships with substitute carers and the degree to which they accept the need to be in care (Biehal et al., 2010; Sinclair et al., 2005). Children in these circumstances may yearn for rejecting or dangerous parents and may sometimes say what they imagine a parent would want to hear. They may feel hurt, angry, guilty or confused. While it is indisputable that children should be consulted, there is a fine line between consulting with children and over-burdening them with responsibility. Fox Harding has pointed to the ambivalence in child liberationist writing as to whether an enlargement of children's rights should be accompanied by an increase in children's responsibilities, suggesting that this ambivalence creates "an unfortunate space in which it is possible for children to have inappropriate responsibility for themselves thrust upon them by default" (Fox Harding, 1996: 148). As Howe has argued, asking children to assume too much responsibility for the quality of their own lives "may be developmentally demanding and psychologically unhelpful" (Howe, 1997: 168).

Children's accounts, and their views of their experiences, must be taken seriously. However, children should not be viewed as consumers of services, making choices between different options on offer. If children are to be empowered to make a meaningful contribution to decisions about their lives, professionals must focus not only on surface rationality but on depth and complexity. Sometimes, a balance may need to be struck between children's rights to have "due weight" given to their views and their rights to "special protection".

4. CONCLUSION: WELL-BEING, POVERTY AND RIGHTS

The model of children's rights specified by the UNCRC offers a helpful balance between protection and autonomy. It calls for absolute protection from harm for all children, while upholding the rights of children to express their views freely in all matters affecting them. This is a helpful framework in which to locate discussions about raising the well-being of children in public care, as it explicitly tackles issues of protection, discrimination and participation, which are of particular relevance to this group. However, questions of participation and autonomy are more complex. For maltreated children and those in public care, there may sometimes be a tension between children's and parents' rights, a tension that the state has dealt with in different ways in different periods. The complexity of the circumstances and relationships of children in public care also highlights the tension between taking children's views seriously and over-burdening them with responsibility for their lives, raising questions about the relative importance of the right to autonomy and the right to protection, and about the balance between these rights.

A rights-based approach throws questions about child poverty into sharp relief.

Irrespective of whether children's rights are conceptualised predominantly in terms of protection or of autonomy, these rights are compromised by poverty. Poverty is known to be associated with poor health, poor educational progress and maltreatment, for example, and generally limits choices and the control that individuals have over their lives (Fox Harding, 1996). Poverty compromises not only children's rights to provision but also their rights to health and development. For children who enter public care, the direct and indirect effects of poverty on their well-being are complex. Before they enter care, poverty is one of a number of inter-related adversities which may compromise their well-being, but not the sole, or principal, cause of their difficulties. On leaving care, they are again at risk of poverty as young adults. A focus on raising the well-being of these children while they are in care may not only have positive consequences for them during the course of their childhood but also reduce the risk of poverty in late adolescence and after, once they have left public care.

REFERENCES

ALDERSON, P. (1992), Rights of children and young people, in: COOTE, A. (ed.) *The welfare of citizens. Developing new social rights.* London: IPPR/Rivers Oram Press.
ANDERSEN, S. & FALLESEN, P. (2010), A question of class: on the heterogeneous relationship between background characteristics and a child's placement risk, *Children and Youth Services Review*, (32): 783–789.
BEBBINGTON, A. & MILES, J. (1989), The background of children who enter local authority care, *British Journal of Social Work*, (19): 349–368.
BEN-ARIEH, A. (2008), The child indicators movement: past, present and future, *Child Indicators Research*, (1): 3–16.

BERRIDGE, D. (2007), Theory and explanation in child welfare: education and looked after children, *Child & Family Social Work*, (12): 1–10.
BERRIDGE, D. & SAUNDERS, H. (2009), The education of fostered and adopted children, in: SCHOFIELD, G.A.S., J. (ed.), *The Child Placement Handbook. Research, Policy and Practice*. London: BAAF.
BIEHAL, N. (2005), Working with adolescents. Supporting families, preventing breakdown. London: BAAF.
BIEHAL, N., CLAYDEN, J., STEIN, M. & WADE, J. (1995), *Moving on. Young people and leaving care schemes*. London: HMSO.
BIEHAL, N., ELLISON, S., BAKER, C. & SINCLAIR, I. (2010), Belonging and permanence. Outcomes in long-term foster care and adoption. London: BAAF.
BIEHAL, N. & SAINSBURY, E. (1991), From values to rights in social work, *British Journal of Social Work*, (21): 245–257.
BRADSHAW, J. & RICHARDSON, D. (2009), An index of child well-being in Europe, *Child Indicators Research*.
BRADSHAW, J., RICHARDSON, D., RITAKALLIO, V.M. (2007), Child poverty and child well-being in Europe, *Journal of Children's Services*, (2): 18–36.
CHILDREN'S RIGHTS ALLIANCE FOR ENGLAND (2009), Lessons for the UK to learn from other countries' implementation of the CRC – why and how we should make the CRC part of our national law. *Presentation to the All Party Parliamentary Group for Children*. London.
DARKER, I., WARD, H. & CAULFIELD, L. (2008), An analysis of offending by young people looked after by local authorities, *Youth Justice*, (8): 134–148.
DEPARTMENT OF HEALTH (2001), Children looked after by local authorities year ending 31 March 2000 England. London: Department of Health.
DEPARTMENT OF HEALTH AND SOCIAL SECURITY (1985), *Review of child care law*. London: HMSO.
DIENER, E. (1984), Subjective well-being. *Psychological Bulletin*, (95): 542–575.
DIXON, J., LEE, J., WADE, J., et al. (2006), *Young people leaving care. A study of costs and outcomes*. York: University of York.
DOYAL, L. & GOUGH, I. (1991), *A theory of human need*. Basingstoke: Macmillan.
FANSHEL, D. & SHINN, E.B. (1978), *Children in foster care. A longitudinal investigation*. New York: Columbia University Press.
FATTORE, T., MASON, J. & WATSON, E. (2009), When children are asked about their well-being: towards a framework for guiding policy, *Child Indicators Research*, (2): 57–77.
FINKELHOR, D. (1994), The international epidemiology of child sexual abuse, *Child Abuse & Neglect*, (18): 409–417.
FISHER, M., MARSH, P. & PHILLIPS, D. (1986), *In and out of care*. Batsford/British Agencies for Adoption and Fostering.
FORD, T., VOSTANIS, P., MELTZER, H., et al. (2007), Psychiatric disorder among British children looked after by local authorities: comparison with children living in private households, *British Journal of Psychiatry*, (190): 319–325.
FORRESTER, D. & HARWIN, J. (2000), Monitoring children's rights globally: can child abuse be measured internationally? *Child Abuse Review*, (9): 427–438.
FOX HARDING, L.M. (1991), *Perspectives in child care policy*. London: Longmans.

FOX HARDING, L.M. (1996), Recent developments in "children's rights": liberation for whom? *Child and Family Social Work,* (1): 141–150.

FRANKLIN, B. (1989), Children's rights: developments and prospects, *Children and Society,* (3): 50–66.

GOLDSTEIN, J., FREUD, A. & SOLNIT, A. (1973), *Beyond the best interests of the child.* New York: Free Press.

GOUGH, I. & MC GREGOR, J.A. (eds.) (2007), *Well-being in developing countries: from theory to research.* Cambridge: Cambridge University Press.

HILL, C.M. (2009), The health of looked after children, in: SCHOFIELD, G. & SIMMONDS, J. (ed.), *The child placement handbook. Research, policy and practice.* London: BAAF.

HOLMAN, B. (1988), Putting families first: prevention and childcare, Macmillan.

HOLT, J. (1975), *Escape from childhood.* Harmondsworth: Penguin.

HOWE, D. (1997), Psychosocial and relationship-based theories for child and family social work: political philosophy, psychology and welfare practice, *Child and Family Social Work,* (2): 161–169.

INNOCENTI RESEARCH CENTRE (2007), Child poverty in perspective: an overview of child well-being in rich countries. Innocenti Report Card 7. Florence: UNICEF Innocenti Research Centre.

JACKSON, S. & MARTIN, P. (1998), Surviving the care system: education and resilience, *Journal of Adolescence,* (21): 569–583.

LINDSAY, M. (1992), An introduction to children's rights. NCB Research Highlight. London: National Children's Bureau.

LINDSAY, M. (1998), Discrimination against young people in care: theory of careism, *Childright,* (151): 11–13.

LOWE, N., MURCH, M., BADER, et al. (2002), *The plan for the child. Adoption or long-term fostering.* London: BAAF.

MALUCCIO, A. & FEIN, E. (1983), Permanency planning: a redefinition, *Child Welfare,* (62): 195–201.

MCCANN, J., JAMES, A., WILSON, S., et al. (1996). Prevalence of psychiatric disorders in young people in the care system, *British Medical Journal,* (313): 5129–5130.

MELTZER, H., GATWARD, R., GOODMAN, R., et al. (2000), *The mental health of children and adolescents in Great Britain.* London: The Stationery Office.

MELTZER, M., GATWARD, R., CORBIN, T., et al. (2003), *The mental health of young people looked after by local authorities in England.* London: The Stationery Office.

MUNRO, E. & WARD, H. (2008), Balancing parents' and very young children's rights in care proceedings: decision-making in the context of the Human Rights Act 1998, *Child and Family Social Work,* (13): 227–234.

PACKMAN, J. (1993), From prevention to partnership: child welfare services across three decades, *Children & Society,* (7): 183–195.

PACKMAN, J. & JORDAN, B. (1991), The Children Act: looking forward, looking back, *British Journal of Social Work,* (21): 315–327.

PARKER, R. (1980), *Caring for separated children.* London: Macmillan.

PARTON, N. (1991), Governing the family. Child care, child protection and the state. London: Macmillan.

PICKETT, K.E. & WILKINSON, R.G. (2007), Child wellbeing and income inequality in rich societies: ecological cross sectional study, *British Medical Journal*, (335): 1080–1085.
PRISON REFORM TRUST (1991), *The identitikit prisoner. Characteristics of the prison population*. London: Prison Reform Trust.
READING, R., BISSELL, S., GOLDHAGEN, J., et al. (2009), Promotion of children's rights and prevention of child maltreatment, *The Lancet*, (373): 332–343.
REES, G., BRADSHAW, J., GOSWAMI, H., et al. (2010), *Understanding children's well-being: a national survey of young people's well-being*. London: The Children's Society [Accessible at www.childrenssociety.org.uk/research].
ROSE, N. (1999), *Inventing ourselves. Psychology, power and personhood*. Cambridge: Cambridge University Press.
ROWE, J. & LAMBERT, L. (1973), *Children Who wait*. London: Association of British Adoption and Fostering Agencies.
ROWLANDS, J. & STATHAM, J. (2009), Numbers of children looked after in England: a historical analysis, *Child and Family Social Work*, (14): 79–89.
RUBIN, D.M., HAFNER, L., LUAN, X., et al. (2009), Welfare reform and beyond working paper. Placement stability and early behavioural outcomes among children in out-of-home care, in: *Child protection: using research to improve policy and practice, 2009*. The Brookings Institution, 1–31.
SECRETARY OF STATE FOR SOCIAL SERVICES (1974), Report of the Committee of Inquiry into the Care and Supervision Provided in Relation to Maria Colwell. London: HMSO.
SIDEBOTHAM, P. & HERON, J. (2006), Child maltreatment in the "children of the nineties": a cohort study of risk factors, *Child Abuse and Neglect*, (30): 497–522.
SINCLAIR, I., BAKER, C., WILSON, et al. (2005), *Foster children. Where they go and how they get on*. London: Jessica Kingsley Publishers.
SOINTU, E. (2005), The Rise of an ideal: tracing changing discourses of wellbeing, *The Sociological Review*, 255–274.
STEIN, M. (2005), Young people leaving care: poverty across the life course, in: PRESTON, G. (ed.), *At greatest risk: the children most likely to be poor*.
STEIN, M. (2009), Young people leaving care, in: SCHOFIELD, G. & SIMMONDS, J. (ed.), *The child placement handbook. Research, policy and practice*. London: BAAF.
STEIN, M. & CAREY, K. (1986), *Leaving care*. Oxford: Blackwell.
THE CHILDREN'S SOCIETY (2006), *Good childhood? A question for our times*. London: The Children's Society.
THOBURN, J. (2009), International contexts and comparisons, in: SCHOFIELD, G. & SIMMONDS, J. (ed.), *The child placement handbook. Research, policy and practice*. London: BAAF.
WADE, J., BIEHAL, N., CLAYDEN, J., et al. (1998), *Going missing: young people absent from care*. Chichester: Wiley.
WARD, H., MUNRO, E.R. & DEARDEN, C. (2006), *Babies and young children in care*. London: Jessica Kingsley Publishers.

6. 'POVERTY, LIKE BEAUTY, LIES IN THE EYE OF THE BEHOLDER'?

Jan VRANKEN

INTRODUCTION

'For deciding who is poor, prayers are more relevant than calculation because *poverty, like beauty, lies in the eye of the beholder*'. Was Mollie Orshansky[1] (1965, 1969) right when she opened one of her many influential contributions on poverty with this statement? Is poverty just a matter of perception and are numbers and other quantitative data no more than illustrations – or, at best, the inevitable result – of how we define poverty, a definition that is itself the logical outcome of a specific perception? Or is any definition, description and analysis of poverty the result of a complex of interactions between data and perceptions?

As for the study of child poverty, other variables enter this already complicated equation. Perhaps the most common complication is that some matters that public (and political) opinion automatically identifies with growing up in poverty – such as child abuse – are present in all layers of society. It is, therefore, important to realise that children in poverty are excluded twice: as persons living in poverty and as persons who have even less control over their life in poverty than others.

There is a fairly common recognition that if poverty and social exclusion in general are to be reduced in the long term, the cycle of child poverty needs to be broken in the short term; it is not only the well-being of children that is important, but also the well-becoming (see Roelen, chapter 7). The other dimension is of equal importance: if we want to break the cycle of child poverty, then the relevant features in the child's present environment have to be tackled systematically. Reviewing the empirical evidence, Wagmiller et al. (Wagmiller and Aber 2006) state that 'compared with children from more economically secure families, poor children tend to perform worse on a wide variety of

---

[1] In 1963 Mollie Orshansky (1915–2006) developed the official measure of poverty that is still used by the US government today.

measures'. They have higher infant mortality rates, more asthma, more physical and mental health problems and worse health generally; they are less well adjusted and report lower self-esteem, their grades are lower; they are more often placed out of age-appropriate regular classrooms and attain fewer years of schooling than their more economically advantaged peers. As adults, children from less affluent families are more likely to be unemployed or underemployed, to earn lower wages when they do work, and to be poor themselves.

In this chapter we intend to focus on the impact of the perspective that is being used – sometimes, that has been chosen – for the study of poverty. There are two reasons for this focus. First, poverty in general constitutes the context for describing, analysing and combating child poverty. Secondly, the importance of the 'perspective' on or the 'model' of poverty that structures whatever is written or said on poverty has always been underestimated, especially in policy-making and research circles.

I will discuss the following items:

1° As a starter: the UNICEF approach
2° Different perspectives on poverty
3° Different poverty models in the social sciences
4° A 'culture of poverty'?
5° From a static to a dynamic (life course) perspective
6° The importance of networks
7° Between the social and the spatial: does social cohesion imply social exclusion?
8° Does a general framework suffice for understanding child poverty?

1. WHAT UNICEF SAYS...

Since UNICEF is undoubtedly the organisation that focuses most on child poverty worldwide, its definition of 'children living in poverty' is a good starting point. According to 'The State of the World's Children Report' (UNICEF, 2005), children are living in poverty when they experience deprivation of the material, spiritual and emotional resources needed to survive, develop and thrive, leaving them unable to enjoy their rights, to achieve their full potential or to participate as full and equal members of society.

UNICEF accepts that 'all practicable definitions of poverty are ultimately definitions of relative poverty' (that is 'related to time and place') and recognises the shortcomings of an income definition, such as that a family's economic resources also depend on savings and home ownership, and that the income definition is related to the country's income distribution. It recognizes that poverty, and especially child poverty, has many other dimensions, such as the quality of the social environment.

However, when it comes to empirical analyses of poverty UNICEF uses an income definition: not the higher EU poverty line (60% of median income),[2] but the lower OECD definition (50% of median income). At the same time, UNICEF calls for further research 'to develop poverty measures that will provide a better guide'. Indeed, traditional poverty measures are not a very appropriate guide because they lack the inspiration of complex definitions of poverty, coherent conceptual and theoretical frameworks and, also, recognition of the importance of the perception of poverty. This means that we need, firstly, even more sophisticated indirect poverty measures (including the diversity of income sources, taking account of depreciation, and using equivalence scales that attribute more weight to children); secondly, a shift to direct and multidimensional measures using educational, housing and health indicators and the like; and thirdly, a coherent and elucidating framework. I will proceed to discuss these different options in the following sections:[3]

Another important matter is that most definitions and sets of indicators of poverty and deprivation have been developed without considering the heterogeneity of the population living in poverty; they only take account of their 'homogeneous' fate – to cite the Dutch sociologist and author Durlacher (Durlacher 1973). So, an important item for further discussion is whether the standard definitions and measures are as relevant for describing and understanding the position and living conditions of children in poverty as they are for the poverty phenomenon in general, an insight that is present in following considerations on the relevance of the EU Laeken indicators:[4]

> *However, one area where consultation and involvement has remained underdeveloped in most Member States is in relation to children and young people. There is little evidence of how, if at all, they have been included in the NAPs process nor is there*

---

[2] This definition also is the one used by the European Committee of Social Rights, the expert committee that judges whether States Parties are in conformity in law and in practice with the provisions of the European Social Charter, the fundamental treaty of the Council of Europe guaranteeing social and economic human rights (see www.coe.int/t/dghl/monitoring/socialcharter/).

[3] The need for a multidimensional measure of poverty does not invalidate income and its distribution as the leading indicator of poverty; it 'remains the most telling single indicator of child wellbeing' and 'a central focus of political and public concern' (UNICEF Innocenti Research Centre, *Child Poverty in Rich Countries*, Report Card No. 6, Florence, Italy: UNICEF Innocenti Research Centre, 2005:7. Available at www.unicef.org/irc.).

[4] This topic is the subject of Keetie Roelens' contribution to this volume.

evidence of linking the NAP process to existing children and young peoples structures such as school, youth parliaments and youth councils. This is surprising given the high priority given in many plans to child poverty.
(From the Joint Report on Social Inclusion, 2005: 114)

2. DIFFERENT PERSPECTIVES

In his seminal work on poverty, Peter Townsend (Townsend 1979) grasped very well the importance of the perspective used, stating that there is a sequence which runs from perspective to definition to analysis to policy-making.

In my previous work, I systematised the existing perspectives on poverty by further developing an older typology (Eames and Goode 1973). In two stages, I have arrived at six perspectives. These perspectives are classified according to the level at which causes are looked for (the individual, institutional or societal level – or the micro-, meso- or macro-level) and to the location of those causes (internal or external: inside or outside the individual, the institution or society at large). The resulting perspectives are – explicitly or implicitly – structuring theoretical debates, empirical research, policy-making, welfare and public opinion and they might be seen – as Townsend did – as connecting these different types of social action.

Table 1. Six models

	Internal cause	External cause
Micro (individual level)	Personal deficiency	Personal accident
Meso (institutional level)	Institutional deficiency	Institutional stigmatisation
Macro (societal level)	Structural	Cyclical

The idea that causes of poverty have to be looked for at the micro-level – that of individuals – still dominates public and political perceptions, explaining social exclusion in terms of personal deficiency, deviant behaviour or accidents. The harshest version is that of 'blaming the victim': the poor and the socially excluded are not only the authors of their own misfortune but even of that of other members of society. It should be clear that this perspective is less applied to those described as the 'deserving poor'; children, widows, handicapped people are not considered to be responsible for their fate. This micro-perspective is very popular with economists and psychologists. Economists introduced uncertainty and erroneous perceptions about the constraints under which utility is maximised, and uncertainty about the consequences of one's decisions, as – at least partial – explanations of social exclusion. This has led to poverty being perceived as the unfortunate result of an unsuccessful attempt to maximise welfare. The 'accident' model has inspired most continental systems of social security.

At the other side of the spectrum we identified two macro perspectives. The cyclical approach refers to phenomena that 'happen' to societies, such as natural catastrophes (earthquakes, tsunamis). Droughts and floods, pandemics, economic crises and rapid social change are often also conceived in terms of phenomena that fall outside human control – which, however, is not the case. The structural perspective, on the other hand, implies that poverty is produced by the way society is organised and by the processes that are generated by this organisation. From this perspective, society's organisation is the problem, not poverty.

This macro-level approach goes further in its structural implications than the institutional approach, which is situated at the perhaps most sociological of all levels: the meso-level. At this institutional level we identified an (institutional) deficiency model, which attributes social exclusion to the physical and social thresholds of a service, and an approach we could call 'social stigmatisation'; the stigma attached to certain social services and – as a consequence – to their clientele/users is a good illustration.

3. DIFFERENT POVERTY MODELS IN THE SOCIAL SCIENCES

Although the six perspectives inspire scientific models, the latter (social theories, hypotheses) transcend their simple form due to their inclusion in what we could call a 'research culture' with its own standards, expectations and social control mechanisms. Academic approaches to poverty are usually a mix of different perspectives, with one of them in a dominant position. This will become clear when we take a closer look at the three main social science frameworks for poverty and social exclusion: the economic welfare model, the capabilities model and the social exclusion framework.

3.1. ECONOMIC WELFARE: FROM A ONE-DIMENSIONAL TO A MULTIDIMENSIONAL CONCEPTUALISATION

'In the beginning, poverty was only a lack of money.' On closer examination, this 'received wisdom' about the genesis of poverty studies needs some modification. The first scholars already defined poverty in larger and more structural terms than became customary about a century later. This tradition stretched from

Engels (the sociologist of the famous couple Karl Marx and Frederick Engels, see his *On the Condition of the Working Class in England*, 1845) to Charles Booth in his seventeen-volume study of *The Life and Labour of the People in London* (1889–1903). The fate of children occupied a central position in their concerns.[5]

This structural approach was gradually replaced by a functionalist one; the study of the condition of society had to make way for the study of specific social problems. Poverty was reformulated in terms of individual behaviour and subculture; the social deviance perspective is the clearest theoretical manifestation of this trend. It went from bad to worse; in the end, almost every conceptual or theoretical consideration disappeared from view. The quasi-monopoly of a strict income definition of poverty diverted attention from its causes, processes, structure and context. It is only when new, non-standard items were included in the income definition that things began to move.

First, the definition of income in terms of cash income was enlarged with other income sources, including fringe benefits and other income in kind. In a certain sense, the redefinition of poverty as a multidimensional phenomenon is the ultimate phase in this process, which is not the subject of this chapter. More relevant is the categorisation of poverty according to its definitions and ensuing measuring methods, which is usually done in terms of dichotomies: absolute

---

[5]   Let us illustrate this with some fragments from Engels' impressive empirical study 'On the Condition of the Working Class in England' (Engels, 1845):
The food of the labourer, indigestible enough in itself, is utterly unfit for young children, and he has neither means nor time to get his children more suitable food. Moreover, the custom of giving children spirits, and even opium, is very general; and these two influences, with the rest of the conditions of life prejudicial to bodily development, give rise to the most diverse affections of the digestive organs, leaving life-long traces behind them. (…) But new disease arises during childhood from impaired digestion. Scrofula is almost universal among the working-class, and scrofulous parents have scrofulous children, especially when the original influences continue in full force to operate upon the inherited tendency of the children. A second consequence of this insufficient bodily nourishment, during the years of growth and development, is rachitis, which is extremely common among the children of the working-class. The hardening of the bones is delayed, the development of the skeleton in general is restricted, and deformities of the legs and spinal column are frequent, in addition to the usual rachitic affections. (…) Children who are half-starved, just when they most need ample and nutritious food – and how many such there are during every crisis and even when trade is at its best – must inevitably become weak, scrofulous and rachitic in a high degree. And that they do become so, their appearance amply shows.
(…) The death-rate is kept so high chiefly by the heavy mortality among young children in the working-class. The tender frame of a child is least able to withstand the unfavourable influences of an inferior lot in life; the neglect to which they are often subjected, when both parents work or one is dead, avenges itself promptly, and no one need wonder that in Manchester, according to the report last quoted, more than fifty-seven per cent of the children of the working-class perish before the fifth year, while but twenty per cent of the children of the higher classes, and not quite thirty-two per cent of the children of all classes in the country die under five years of age.

versus relative poverty, objective versus subjective poverty, material versus immaterial poverty, direct versus indirect poverty measures.

The core of the first dichotomy is that absolute poverty refers to a substantive approach of the phenomenon and usually leads to a 'subsistence' definition of poverty. Although this is not always the case, as is shown by Sen's capabilities approach[6]; when it comes to applying such an 'absolute' definition of poverty in specific societal contexts, some degree of relativity is almost inescapable. Relative standards are based on this idea, that poverty is a contextual phenomenon and thus any study has to take into account its historical, spatial and cultural context. Because of this linking of poverty to varying circumstances, a relative definition of poverty contains a self-updating mechanism. In very practical terms, this means that any change in the average living standard (as expressed in income terms or in terms of a multiple index) of the reference society automatically leads to a corresponding change in the definition of poverty. We could even identify a more extreme form of a relative poverty definition, i.e. a relational one, which takes the relation between groups and persons in poverty and those out of poverty as its core subject or the link between conditions of poverty and societal structures and processes. These may then range from specific subsystems such as the labour market, the education system or health care to all-encompassing realities such as (late, neo-) capitalist or risk society.

The objective/subjective dichotomy has a slightly different meaning, depending on whether it refers to the data used or the actors defining the poverty line. Traditionally, objective stands for 'external (non-poor) experts defining what poverty is and how to measure it'; subjective poverty is poverty as experienced by people living in poverty. Objective may, however, also refer to the use of external data, subjective to the experiences of people living in poverty. Because these data will often be obtained through surveys, they will also obtain the status of 'objective' in the sense of 'exceeding the level of pure impressions'. Which means that external experts will base their definition of poverty on the 'subjective' feelings of the poor, an approach that was fairly popular in the 1970s, as is well illustrated by the early poverty definitions of the Antwerp Centre for Social Policy (1976 and later) and the Leyden Poverty Line (LPL). This method is at present better known as the consensual method, based on survey data giving the evaluation by respondents of their own income situation and/or the income deemed necessary to make ends meet.

What about the set 'direct' versus 'indirect' income measures? Direct income definitions measure poverty in terms of outcomes, and are based on indicators of deprivation such as for health, education or housing. Indirect poverty measures

---

[6] See later in this chapter.

express poverty in terms of opportunities: resources that persons and households have at their disposal and with which they may 'buy' the social goods that are available for them. They are usually based on income and the ensuing income poverty line can be constructed in several ways: as expert budgets, as a political-administrative poverty line, in terms of consensual methods (see above) and/or as a relative income threshold.

Expert budgets are based on a basket of goods and services that are deemed necessary to achieve a certain living standard; the total cost of such a basket is the poverty line. They are the oldest form of poverty measure because they go back to Booth (Booth 1989) and Rowntree (Rowntree 1902). The official US poverty thresholds still are based on such budgets and in the UK, Bradshaw (Bradshaw 1993) is the best known protagonist of this approach.[7] This implies that Booth, Rowntree, Bradshaw and the like are the experts concerned, which leads to two important questions. The first is: 'Are the criteria that the experts use so objective?'; the second: 'Are researchers the only persons we could call experts'?[8]

A substantial part of our discussion boils down to the *hypothesis* that all poverty lines are arbitrary because they are all based on some value judgment. A typical aside in this respect is: 'Is it acceptable to include alcohol, tobacco, newspapers or the occasional cup of coffee in a café in this basket or is that going too far?' This questions the monopoly of researchers as experts, and of 'objective' criteria (based on research) as the only valid ones for defining poverty measures. Are not practical experience (living in poverty) or political concerns (trying to avoid conflicts between different systems of social protection, such as between social security and social assistance) equally valid? A positive answer to this question would put policy-makers and the poor themselves into the picture as 'experts' and would qualify administrative thresholds or 'institutional' poverty lines, such as the guaranteed minimum income (social assistance), as expert definitions. Indeed, such poverty lines are an indication of how much society wishes to spend on providing a minimum standard of living for its 'weakest' members. In the words of the famous doctor Samuel Johnson, 'a decent provision for the poor is the true test of civilization' (Boswell 1791).

There are, however, other interesting features present here which uncover the multiplicity of any poverty definition or poverty line, such as that it is liable to manipulation: the lower the threshold, the lower the number of poor. If such a

---

[7] This approach was also used for the first serious estimation of the number of poor in Belgium in the late 1960s (Vranken, 1971).

[8] We do not discuss in this chapter the technical problems related to the use of this approach, such as how to update the budget, the taking into account of geographical and cultural differences, or the difficulties with comparative research.

threshold is used as the official definition of poverty, the more public authorities spend on poverty, the more poor people there are in a society.[9]

3.2. THE CAPABILITIES APPROACH

Over the last decade Amartya Sen's capability approach has become the leading alternative to standard economic frameworks for thinking about poverty and inequality. Sen emphasizes that poverty must be seen as the deprivation of basic capabilities rather than merely as low income. Some dimensions of wellbeing of people are not easily captured by monetary indicators. In his many articles and books, he has developed, refined and defended a framework that considers *poverty as deprivation of basic capabilities*. A *capability* is 'the freedom to achieve valuable beings and doings' and 'represents the various combinations of functionings (beings and doings) that the person can achieve'. *Functionings* are the various things a person may value doing or being and are *constitutive* of a person's being. Achieved functionings are measurable, observable and comparable, such as literacy and life expectancy.

Capability is, thus, a set of vectors of functionings, reflecting the person's freedom to lead one type of life or another to choose from possible livings. Both freedom and capabilities have an intrinsic and instrumental value.

The two central elements in *functionings* are valuable beings and doings. These are all the 'ends' of human life, although they can also be means. They can be elementary (escaping morbidity and mortality; nourishment; mobility) or complex (self-respect; participation in community life; ability to appear in public without shame); general (capability to be nourished) or specific (capability to drink clean water). The specific form that their fulfilments may take would tend to vary from society to society.

For Sen (Sen 1999), development is 'a process of expanding the real freedoms that people enjoy' and its objective is the promotion and expansion of valuable capabilities. Freedom, for Sen, has two aspects: a process and an opportunity aspect. The process aspect refers to the ability to be agents – to affect the processes at work in their own lives or as general rules in the working of society. The opportunity aspect is about the *ability to achieve* valued functionings.

How is social exclusion to be defined in this context? The *process* of social exclusion produces a *state* of exclusion that can be interpreted as a combination

---

[9] Again, some technical decisions contain value judgments; for example, if the household is used as the unit of account, an equal distribution of means within the household is supposed.

of relevant deprivations. So, according to Sen's capability approach, social exclusion can be understood as the impossibility to achieve some relevant functionings leading to a state of deprivation. For this definition to become operational "the relevant functionings" need to be identified as well as the excluded individuals in every dimension and their degree of exclusion.

A key dilemma for the capabilities approach has been how to measure what people could do, as opposed to what they actually do. Capability indicators can be found in standard datasets and more significantly, it is possible to develop new survey instruments which operationalise the list of capabilities. The main capabilities measurement instrument has over 60 indicators, is being used by a number of research groups and is discussed in Anan et al. (Anand, Santos et al. 2009).

3.3. POVERTY AS A FORM OF SOCIAL EXCLUSION

Halfway through the 1990s, exclusion became the paradigm within which society reflected upon itself and its dysfunctions, and sometimes also tried to find solutions (Paugam 1996). The increased use of the notion of social exclusion then led to a proliferation of meanings (Vranken, Geldof et al. 1997; Vranken 1998). One definition of social exclusion is couched in the Anglo-Saxon tradition of *citizenship*. From this perspective, social exclusion is described 'in terms of denial or non-realization of *social rights*' (Room, Berghman et al. 1991), such as 'the right to a certain standard of living' or, more specifically, the right to work, to housing and to education. Social exclusion, then, implies that access to these rights is, knowingly or unknowingly, restricted by the manner in which social services are organised, or by the vulnerable economic, social and political position of certain citizens. The two other definitions are continental, i.e. rather French. According to one view, the notion of social exclusion concerns the *gap* that exists between situations or groups in one or more areas of social life; this gap need not necessarily relate to poverty. According to the other view, social exclusion is a *process*, while poverty is one of its results.

In the end, however, in the EU jargon exclusion was viewed as the 'modern', 'multidimensional' form of poverty, reducing poverty (again) to its sole income dimension, thus throwing overboard two decades of important empirical, conceptual and theoretical research. This consideration led me to redefine the relationship between social exclusion and poverty, in that I considered poverty as one of many forms of social exclusion, and that I used social exclusion as a generic concept, at the same level as three other forms of difference in society: inequality, differentiation and fragmentation (Vranken 2001).

This typology is based on two variables: the presence or absence of a hierarchical relation between social units and the presence or absence of a fault line (rupture, 'sudden discontinuity') between them. Fault lines were defined at the micro level (relational fault lines in social networks), the meso level (exclusion from institutions or isolation in spatial entities), and the macro level (sub-societies, as in the dual labour market, or different cultures).

Table 2

	Hierarchy	
Fault lines, ruptures	no	yes
No	differentiation	inequality
Yes	fragmentation	exclusion

Source: Vranken, 2001.

Equivalent social relations lead to social differentiation, while subordinate relations result in social inequality. Social differentiation refers to different tastes in food or clothing, colour of eyes or hair and other bodily differences, as long as these do not have a notable impact on the distribution of important social goods (income, status, power). When the latter is the case, we have social inequality. In both cases, there is no fault line between the units under survey. Social fragmentation refers to different fields that are clearly separated from each other (by fault lines) but which are not subordinated mutually, as in a polycentric (social or spatial) area with equivalent fragments (ethnic villages or a 'real' multicultural society).

Social exclusion implies both a hierarchical relationship between individuals, positions or groups and their separation by clearly discernible fault lines. Certain fault lines result from collective intervention (e.g. the establishment of a subsistence income, institutional isolation, absence of political or social rights), while others arise from societal structures (racism, segmented labour markets). 'Social exclusion' is thus a generic concept that refers to various situations and processes such as polarisation, discrimination, poverty and inaccessibility.

Poverty, then, is a special case of social exclusion:[10] it is an accumulation of interrelated forms of exclusion.[11] Poverty refers to non-participation or a very limited participation in various social commodities such as income, labour,

---

[10] Expressed in terms of a conceptual equation: Poverty = $\int \Sigma$ (social exclusion a + social exclusion b + ... social exclusion z)} + $\int$interaction between a, b, c...z}. Or: Poverty = social inequality + fault line + interrelated fields.

[11] Social exclusion is used as shorthand for multifarious quantitative peripheral situations, including forms of inequality. Those may accumulate to such a degree that together they make a qualitative difference (exclusion).

education, housing, health, administration, justice, public services and culture, or to a high degree of participation in features that in themselves are signs of exclusion (or are perceived as such): imprisonment or social assistance. That those areas are multiple and interrelated is what makes poverty a special case of social exclusion, different from homelessness or discrimination. The incapacity of the poor to bridge this complex fault line on their own underlines how powerful a form of exclusion poverty really is.

This definition of poverty (Vranken 1992) includes all those features:

> *Poverty is a network of instances of social exclusion that stretches across several areas of individual and collective existence. It separates the poor from society's generally accepted patterns of life. They are unable to bridge this gap on their own.*

Although this definition of poverty and the implied concept of social exclusion – with its structural connotations – has become a standard reference in Flanders and Belgium alike, the social rights perspective has structured the important General Report on Poverty, which was commissioned by the government and published in 1994. This report was the result of two years of work and brought together organisations in which the poorest had their say along with local public welfare agencies, social workers and many other actors, including labour unions. The main concerns were organised under headings beginning 'The right to …', such as the right to a family, to work, to health, to decent housing, to knowledge and culture. This approach was certainly related to the at that time ongoing (2 February 1994) introduction of social rights into the Belgian Constitution (art. 23).

4. IS THERE A CULTURE OF POVERTY?

Traditional poverty studies inform us about visible characteristics of poverty, but little research is undertaken about 'the structures of daily life of the poor'. We need a 'geology of poverty', to complement the dominant 'geography of poverty'. A qualitative approach is appropriate for reaching people at the fringes of society. Qualitative research can show the reality behind the figures, reveal the complexity of daily life and uncover the deeper roots of poverty.

How people try to cope with situations of deprivation and stress has been the subject of one of the few developed theoretical frameworks in sociology, that of the 'culture of poverty'. It is based on the distinction introduced by Merton (Merton 1968) between culturally defined goals and socially accepted means. Some authors maintain that the poor have a culture of their own, a 'culture of poverty', which enables them to survive in poverty but at the same time prevents them from using resources if and when they become available. Others stick to the hypothesis that the poor just adapt their behaviour to changing circumstances, without values and norms being involved. The most plausible hypothesis is that of an 'adaptational' approach such as that developed by Rodman (Rodman 1963; Rodman 1971) in what he termed the 'lower class value stretch', expressing a dynamic relationship between general values and specific living conditions and scarce resources. Poor people's actual behaviour and behavioural patterns can be explained by a process in which the general values, norms and goals are temporarily 'stretched' – that is adapted to particular circumstances – so as to enable the poor to survive without losing contact with the dominant cultural pattern. It is in this framework that 'coping strategies' also have to be understood.

An important intermediate variable should be taken into account, however, and that is past experience, as expressed in the difference between 'generational' poor and 'new' poor. This variable might preliminarily and roughly be defined as 'early confrontation with exclusion'. At the one end of the continuum, we find people born and raised in poor and deprived families, the so-called generational poor. Their life history is an illustration of early vulnerability in different areas, where a single event or a combination of negative experiences ignites a downward trajectory towards exclusion and poverty. This early confrontation with multiple forms of deprivation has of course major consequences: both personality and basic skills are marked by these experiences and by a constant struggle to deal with inherited vulnerability. Generational poor also encounter opportunities and take initiatives, but in general they do not succeed in permanently reversing their situation because problems keep arising and trying their capacities to cope with them. Periods of harmony are only temporary, and one single negative event can easily tear apart the delicate balance that has developed. It is clear that growing up in poverty seriously clouds one's very future.

At the other end of the continuum, we find people who grew up in a relatively stable and problem-free context. This does not mean that in their trajectories no single trace of bad luck can be found. However, in one way or another they managed to cope with it. They possess enough strength and self-confidence, enough power to deal with the difficulties in their lives. One single setback would generally not bring them into a situation of need, or only temporarily so. The fact that they have ultimately reached a point of deprivation has to be attributed to an accumulation of events. The situation of this group is mostly linked to exclusion from the labour market, itself resulting from poor education.

People who grew up in poverty do not seem to have different expectations about the future from those who grew up with more opportunities. They cherish the same hopes and dreams as middle-class people: a stable job, a caring partner, a nice place to live, a better future for the children. Some succeed in realising their hopes; it is possible to get out of poverty. This confirms that only part of the reality of poverty is captured with cross-sectional analyses; a life course perspective is needed to broaden and to deepen our knowledge about poverty.

5. FROM A STATIC TO A DYNAMIC (LIFE COURSE) PERSPECTIVE

Cross-sectional poverty figures hide a complex reality. Some people are 'accidentally' hit by poverty, some are occasionally poor, some repeatedly move into and out of poverty, and a small group of people are poor over the long term.

> *We thus see that the 7230 persons shown by this inquiry to be in a state of 'primary' poverty, represent merely that section who happened to be in one of these poverty periods at the time the inquiry was made. Many of these will, in course of time, pass on into a period of comparative prosperity (...) But their places below the poverty line will be taken by others who are at present living in that prosperous period...*
> *The proportion of the community who at one period or other of their lives suffer from poverty to the point of physical privation is therefore much greater, and the injurious effects of such a condition are much more widespread than would appear from a consideration of the number who can be shown to be below the poverty line at any given moment (Rowntree 1902): 171–172).*

It is important to study dynamics because of the complex interplay of movements into and out of poverty; that was already clear to Rowntree over a century ago. A 10 percent poverty rate for 10 years may hide two very contrasting realities for the people living in poverty. This poverty figure might result from 1 out of 10 individuals being long-term poor – that is, for 10 years – which means no mobility. However, if every individual in the population is in poverty only for the short term (1 out of 10 years), there is complete mobility.

This has important implications. Poverty is not a 'personal' characteristic, but a position people are allocated to. The poor are just people living in the poorest rooms of the hotel that is our society; some will spent their whole life in that insalubrious room, others will be able to move to a better one – only to return to their previous residence after some time? – and a minority will permanently settle in a room with a view. Identifying the relevant factors in this longitudinal dimension of poverty invites more research into determinants of poverty and more attention to its policy implications. Are specific policy measures needed for the long-term and for the short-term poor? Preventing entry into poverty and stimulating exit out of poverty becomes more important. Indeed, the exit probability decreases as people remain poor for a longer period and this may be explained by a 'scarring effect'.

Determinants of poverty dynamics are to be found at different levels. At the individual level, there is the impact of 'life events'. At the institutional level, differences in mobility rate and poverty profiles will both determine entry and exit rates. There is a link with welfare state typologies. Let us review a selection of important life events, from which the relevance of this approach for child poverty will be quite clear (Dewilde 2003).

Partnership dissolution has a positive and strongly significant effect on the probability of becoming poor. However, the effects are gender-specific: the economic burden of partnership dissolution mainly falls on the shoulders of women – and the children who live with them. The 'positive' effect for men might be overstated, depending on alimony and maintenance payments deductible from the total household income. Having a full-time job offers adequate protection against becoming poor upon partnership dissolution. The effects of *widowhood* on the chances of entering poverty negatively impact on the financial situation of both men and women.

While the pooling of resources is generally considered to be a route out of poverty, *partnership formation* seems to have a positive effect on the poverty entry chance for women. Further analyses suggest that this effect mainly involves women who head a lone parent family in wave (t) and form a household with an economically inactive partner (mostly unemployed) between wave (t) and wave (t + 1).

The *birth of a child* has a positive effect on poverty entry probabilities, but depends on the value of the 'child benefit package' (Ditch, Barnes et al. 1998). This indicator calculates the extent to which welfare states assume responsibility for the cost of raising children. *Leaving the parental home* in order to establish a new household, alone or with a partner or friends, may have a positive effect on the entry probabilities into income poverty, depending on the quality of labour market participation.

In contrast to the situation in Rowntree's days, the *departure of an adult child* has no effect on the poverty entry probabilities for the respondents in the parental household. A last demographic event refers to *'other' changes in the household composition*, such as a grandparent who joins the nuclear unit or a lone mother who moves in with siblings.

What is the impact of the different types of labour market event on the risk of entering income poverty? The household reference person becoming *unemployed or disabled* has a positive and highly significant effect on the poverty entry probabilities. *The transition from a full-time to a part-time job* does not necessarily lead to higher poverty entry probabilities. The negative effects for the partner's transition points to a selection effect: only households that can afford it opt to reduce the female partner's labour market participation. *(Early) retirement* of the household reference person has a positive effect on the poverty entry risks.

6. THE IMPORTANCE OF NETWORKS

Where relational fault lines are at stake, social exclusion may be seen as the result of non-participation in the exchange of social commodities and, by extension, in social networks. Social networks are relational and thus structural par excellence. They refer to a structured multiplicity of social links that define a person's position in society and they usually consist of a mix of strong and weak ties. Weak ties give access to important 'social goods' (labour, income, education), whereas strong ties provide emotional support. For average citizens, both types of tie are found in their personal network; for persons living at the fringes of society, however, this is not so. In their networks, strong ties dominate. However, in order to connect with society's resources – and so to improve their individual or collective position – they need weak ties too.

'Gatekeepers' – policy makers or administrators are important ones – play an important role in preventing poor persons' access to those resources. Because gatekeepers occupy strategic positions in the networks, they have the power to decide whether or not to allow the flow of social commodities to go through. Through providing certain people or groups (the lowly qualified, the immigrants, the homeless) with access to social commodities (such as employment, housing, education, income, status, power) they can promote social inclusion.

Usually, in literature on poverty and social exclusion the supportive functions of networks are highlighted. Strengthening the networks of the poor would increase their opportunities to fully participate in relevant sectors of society, such as the labour market, education and health care. Inherent restrictions on networks are often overlooked, although they help us to understand exactly social exclusion and poverty. One of them is that the 'inclusive' function of networks sometimes turns sour, into 'enclosing' people in their present position. A study of the social mobility of the generational poor ('Bruggen over woelig water' – 'Bridges over troubled water' (Thys, De Raedemaecker et al. 2004) showed that successful upward mobility depended very much on the presence of both an instrumental and an expressive dimension. If only the former (a job, education, a new relation) was present but the latter (integration into the networks of the non-poor and emotional support) stayed behind, the social climbers were doomed to return to their original position. This was also their fate if they did not succeed in cutting off close ties with their former network, including their family of origin. Often this meant that they found themselves in the position of 'marginal persons': in between two stools.

Since networks are often spatially located, e.g. the social networks in a deprived neighbourhood in a specific city, their study reveals the importance of this spatial dimension and brings us closer to the integration of both the social and the spatial dimension of poverty.

7. BETWEEN THE SOCIAL AND THE SPATIAL: DOES SOCIAL COHESION IMPLY SOCIAL EXCLUSION?

Whereas at the level of a society, old and new forms of inequality, exclusion and fragmentation usually exist in a diffuse manner, at the city level and especially in 'poor' or 'deprived' areas, they are often present in such a concentrated way that they capture the imagination (Bourdieu 1993). Poverty and other forms of social exclusion (homelessness, beggars), an increasing concentration of marginalised groups and ensuing spatial segregation constitute the most visible facets of a larger set of problems which also include growing unemployment, increasing criminality, deteriorating private dwellings and public space, dwindling social cohesion, a declining quality of life in some neighbourhoods, and last but not least, the rise of monofunctional areas.

There is an inherent and re-emerging spatial dimension in many forms of social exclusion. Neighbourhoods no longer fulfil the 'integrative' and 'socialising' function as transition zones they still possessed in Burgess's typology (Park, Burgess 1925/1967), but have often become dead-end areas. Moreover, identification with smaller units (neighbourhoods, communities) stands in the way of identification with larger entities (cities, society). The question is, then, whether developments such as increasing fragmentation are obstacles to cohesion and solidarity or rather take away obstacles that formerly prevented identification with more 'modern' – because individualised – levels of social life (the city, the nation state).

In order to have social cohesion – which supposedly is a basic requirement for the sustainability of social units – two conditions must be fulfilled: a 'negative' and a 'positive' one. The degree to which members are willing to participate and identify with a group largely depends upon the degree of internalisation of values and norms and the sanctioning capacities of the group. If people feel that they are part of a group, they are inclined to follow its rules. At the same time, group solidarity is promoted by a common experience of opposition to 'others', of identification with the in-group to such a degree that the out-group is perceived as a 'threat', which leads to the generation of stereotypes and to stigmatisation. This may mean that members are under pressure not to show too much ambition to leave the group and to integrate into 'larger society'. This 'construction of cohesion' is often strengthened and made more visible by its spatial dimension. If part of the group identity is based on feelings of belonging to a certain space, no-go areas and feelings of insecurity for non-group members entering this space are generated.

This supports Healey's (Healey 1998) argument that cohesion and exclusion are not opposing phenomena, but that they imply each other. A strong cohesion may exclude inhabitants from opportunities outside the group (community, neighbourhood). Social cohesion also has a spatial dimension: a high degree of social cohesion within a neighbourhood (strong bonds between the inhabitants of a neighbourhood) may lead to a low degree of social cohesion on the city level (inhabitants of one neighbourhood are not interested in those living in other neighbourhoods) (Vranken 2004). Strong ties between people within communities may lead to social, racial and religious conflicts between these communities and those who are perceived as outsiders. Social cohesion can thus easily breed intolerance. It means that if socially and ethnically diverse groups concentrate in certain areas, their internal cohesion certainly will be fostered but at the expense of their integration at a higher level, as it will also increase the risk of exclusion both for individuals from those highly cohesive communities and of these communities from the rest of society. However, if non-conflicting relations between these diverse groups could be structured at lower spatial levels (neighbourhood or district), a high social cohesion is possible in the urban system as a whole (Vranken 2004).

These different forms of fragmentation threaten solidarity and social cohesion. This is reflected in the withering away of common value patterns, socialising institutions and mechanisms of social control, of neighbourhoods and traditional social classes.

8. IS THIS A RELEVANT FRAMEWORK FOR UNDERSTANDING CHILD POVERTY?

The preliminary question is of course: is this a relevant framework to improve our understanding of poverty in general? Let us suppose that this is the case, since it has inspired many successful research projects and policy proposals. Two questions then remain. Is it specific enough to take account of the specific needs of children? Does it sufficiently consider the relations between children in poverty and the rest of society, so that there is no threat to treat child poverty as an isolated problem?

We intend to take up some of the conceptual and theoretical issues raised in this chapter and have a look at their relevance for a better understanding of child poverty.

Firstly, how important is perception with respect to child poverty? Of utmost importance, especially since our views on child poverty are often clouded by feelings of compassion. The positive side to this is that we never blame children for their poverty; they have always been very 'deserving' poor, which means deserving charity and public assistance. The negative side, however, is that we tend to see child poverty as an isolated problem, because it is so pitiful to see children in poverty but also because it is much easier not to bring in the situations in which those children have grown up and are growing up. If their family situation is taken into account, then it is usually to blame the parent – they provide a screen between the child and society (i.e., the non-poor) onto which responsibility for their poverty can be projected. The institutional and structural approaches, on the other hand, inform us about the context within which child poverty has to be understood. The 'societal organisation' is important for understanding the position of children in economic, cultural and legal terms. Is child labour needed for the survival of the family or does society get more profit from increasing its human capital through compulsory schooling? Are children viewed and treated as young adults or is there a specific status for children? Does legislation protect them against abuse by more powerful adults?

As for the models used in poverty studies, they all provide us with means to measure, to describe, to analyse and to understand child poverty. That the strict income definition of poverty is less useful for informing us about the multifaceted character of child poverty than a wider set of indicators does not need further comment. There is, however, more: if such a set does not contain (enough) child-specific indicators, it does not constitute real progress compared to financial definitions of poverty. Keetie Roelen discusses this dimension in more depth in chapter 7. Also important is the need to collect information on children's perception of their living conditions. This has become the subject of some interesting research projects; see Isabelle Pannecoucke's in chapter 10.

Sen's capability approach is relevant to our subject if only because of its focus on … capabilities. Few concepts are so closely related to the position of children as 'the freedom to achieve valuable beings and doings', because they still have their whole life to realise their capabilities. That is, on condition that they are given at least the ability to be the agents of their own lives.

Both in its rights approach as in its more structural definition, social exclusion helps us to understand better what child poverty is all about. Children in poverty are separated from the rest of society by a social (and sometimes spatial) chasm that they are unable to cross by themselves; even with 'some help from their friends' they will not succeed – on the contrary. Their peer groups, their families, their social networks, their communities and the neighbourhoods they live in are as much excluded as the children are themselves. It is the situation I referred to as 'double exclusion' at the very beginning of my contribution. This exclusion concerns all the domains that are customarily summed up under 'social rights' or 'children's rights'. The structural definition also points our attention towards the several levels in this exclusion from rights, namely that it is not only their presence in the constitution we should be concerned about, but also their translation into legislation, rules and procedures and the accessibility of the institutions concerned with their implementation and control (child care, youth care, schools, sports centres, youth organisations, social housing corporations, health centres).

If children grow up in some 'culture of poverty', which often implies a specific network and a delineated area in the city (a deprived neighbourhood of the dead-end type; transition zones still provide opportunities for spatial and social mobility), it is even more difficult to escape poverty than if they are living amongst middle-class families. This is about (the lack of) role models, about (not) knowing that there is an alternative to poverty, about (not) breaking the intergenerational cycle of poverty. This does not mean that we take the family or the neighbourhood as responsible for the poverty in those children's later lives; that much stronger forces are at work here should have become clear from this contribution. For policy-makers, it is important to identify the life events during which the determinants of poverty we identified play a crucial role; this could provide a clue for effective policy interventions within a structural framework that incites us to stay humble when it comes to assessing the impact of policy-making.

REFERENCES

ANAND, P., C. SANTOS, et al. (2009), in: K. BASU & R. KANBUR, *The Measurement of Capabilities. Arguments for a Better World: Essays in Honor of Amartya Sen.* Oxford: Oxford University Press.
BOOTH, C. (1989), *Life and labour of the people in London.* London: Macmillan.
BOSWELL, J. (1791), *The life of Samuel Johnson.* London: Henry Baldwin.

BOURDIEU, P. (1993), *La misère du monde*. Paris: Seuil.
BRADSHAW, J. (1993), Rediscovering budget standard. The European Face of Social Security, in: J. BERGHMAN & B. CANTILLON, *Essays in Honour of Herman Deleeck*, Aldershot: Avebury: 60–76.
DEWILDE, C. (2003), "A Life-Course Perspective on Social Exclusion and Poverty." *British Journal of Sociology* 54(1): 109–128.
DITCH, J., H. BARNES, et al. (1998), *A Synthesis of National Family Policies 1996*. York: European Observatory on National Family Policies (European Commission).
DURLACHER, G. (1973), "Armoede: een poging tot analyse." *Mens en Onderneming* 27(1): 45–65.
EAMES, E. and J. G. Goode (1973), *Urban poverty in a cross-cultural context*. New York: Free Press.
HEALEY, P. (1998), Institutionalist theory, social exclusion and governance, in: A. MADANIPOUR, G. CARS & J. ALLEN, *Social Exclusion in European Cities: Processes, Experiences and Responses*. London: Jessica Kingsley: 53–74.
MERTON, R. K. (1968), *Social theory and social structure*. New York: The Free Press.
PARK, R. and E. BURGESS (1925/1967), *The City*. Chicago: The University of Chicago Press.
PAUGAM S., Ed. (1996), *L'exclusion: l'état des savoirs*. Paris: La Découverte.
RODMAN, H. (1963), "The lower class value stretch." *Social Forces* 42(2): 205–215.
RODMAN, H. (1971), *Lower-class families. The Culture of Poverty in Negro Trinidad*. New York: Oxford University Press.
ROOM, G., J. BERGHMAN, et al., Eds. (1991). *National policies to combat social exclusion. First Annual Report of the European Community Observatory*. Bath: Centre for Research in European Social and Employment Policy.
ROWNTREE, B. S. (1902), *The Poverty Line. Poverty. A Study of Town Life*. London, Edinburgh, Dublin & New York: Thomas Nelson & Sons: 117–151.
SEN, A., Ed. (1999), *Development as Freedom*. Cambridge: Cambridge University Press.
THYS, R., DE RAEDEMAECKER, W. et al. (2004), *Bruggen over woelig water. Is het mogelijk om uit de generatie-armoede te geraken?* Leuven / Voorburg: Acco.
TOWNSEND, P. (1979), *Poverty in the United Kingdom. A Survey of Household Resources and Standards of Living*. Harmondsworth: Penguin Books.
VRANKEN, J. (1971), "Armoede in België: één op tien." *De Nieuwe Maand* 14(8): 446–460.
VRANKEN, J. (1998), *Social Exclusion: One Player in a Quartet. Paper prepared for the Panel on Poverty Research*, World Congress of Sociology. Montreal: 14.
VRANKEN, J. (2001), Unravelling the social strands of poverty: differentiation, fragmentation, inequality, and exclusion, in: H.T. ANDERSEN & R. VAN KEMPEN, *Governing European Cities. Social Fragmentation, Social Exclusion and Urban Governance*. Aldershot: Ashgate: 71–90.
VRANKEN, J. (2004), Changing Forms of Solidarity: Urban Development Programs in Europe, in: Y. KAZEPOV, *Cities of Europe. Changing Contexts, Local Arrangements, and the Challenge to Urban Cohesion*. Oxford: Blackwell Publishing: 255–276.
VRANKEN, J. & D. GELDOF (1992), *Armoede en sociale uitsluiting. Jaarboek 1991*. Leuven/Amersfoort: Acco.

VRANKEN, J., D. GELDOF et al. (1997), Sociale uitsluiting: deel van een kwartet, in: J. VRANKEN, D. GELDOF & G. VAN MENXEL, *Armoede en sociale uitsluiting. Jaarboek 1997.* Leuven/Amersfoort: Acco: 303–320.

WAGMILLER, J., L. ROBERT, M. C. LENNON, et al. (2006), "The Dynamics of Economic Disadvantage and Children's Life Chances." *American Sociological Review* 71(5): 847–866.

7. CHILD POVERTY – WHAT'S IN A WORD?

Keetie Roelen

INTRODUCTION

Poverty is an undesirable and, to many, unacceptable phenomenon. A wide range of studies has suggested that the hardship of poverty is even more undesirable when it concerns children due to its far-reaching short-term and long-run negative implications (see e.g. Haveman & Wolfe, 1995; Brooks-Gunn and Duncan, 1997; Duncan and Brooks-Gunn, 1997; Esping-Andersen and Sarasa, 2002). The increased acknowledgement that children deserve a special focus in the debate on poverty has encouraged the development of child-specific poverty approaches and measures (see Roelen & Gassmann, 2008; Roelen et al., 2009a).

Despite the widespread consensus that child poverty is an issue that requires more focused attention and analysis, there is little agreement on how to define and measure it (Laderchi et al., 2003). The differences between child poverty approaches in terms of definition, concept and methodology make transparency and clarity about the underlying construct extremely important (see Roelen et al., 2009a; Vandivere & McPhee, 2008; Laderchi et al., 2003). Differences between child poverty approaches, as well as for poverty in general, are a result of theoretical considerations and value judgements (Ravallion, 1994; Vranken, 2010) that lead to different interpretations of reality (Laderchi et al., 2003) and are reflected in diverging poverty outcomes (Saunders, 2004). A lack of transparency or understanding of the definitions, concepts and methodological choices underlying child poverty outcomes can cause misinterpretation of those outcomes and result in inappropriate and inadequate policy responses (Alkire, 2008; Roelen et al., 2009a).

This chapter focuses specifically on the conceptual choices and considerations inherent to child poverty measurement and the consequences of these on empirical outcomes. As such, it aims to show that the discussion of conceptual underpinnings and definitions of child poverty is interesting and important not merely from a theoretical or philosophical point of view, but also from a practical and policy-oriented perspective. A clear understanding of the meaning of the term "child poverty", its origin and its empirical outcomes is important for academics, policy makers and practitioners alike as a lack thereof can lead to inadequate or inappropriate use of the child poverty measure to hand.

Roelen et al (2009a) have developed a generic construction process, which can serve as a framework for the development and interpretation of child poverty approaches and help to overcome problems concerning implicit or unclear decision-making. We will employ this framework to demonstrate the channels that contribute to specific conceptual considerations and the channels through which these considerations impact on empirical outcomes. The framework is derived from a review of construction procedures of existing child poverty approaches and presents a step-wise process to ensure an explicit and solid development of child poverty approaches (Roelen et al., 2009a). The generic construction process and its various steps are illustrated in Figure 1.

Figure 1. Generic construction process

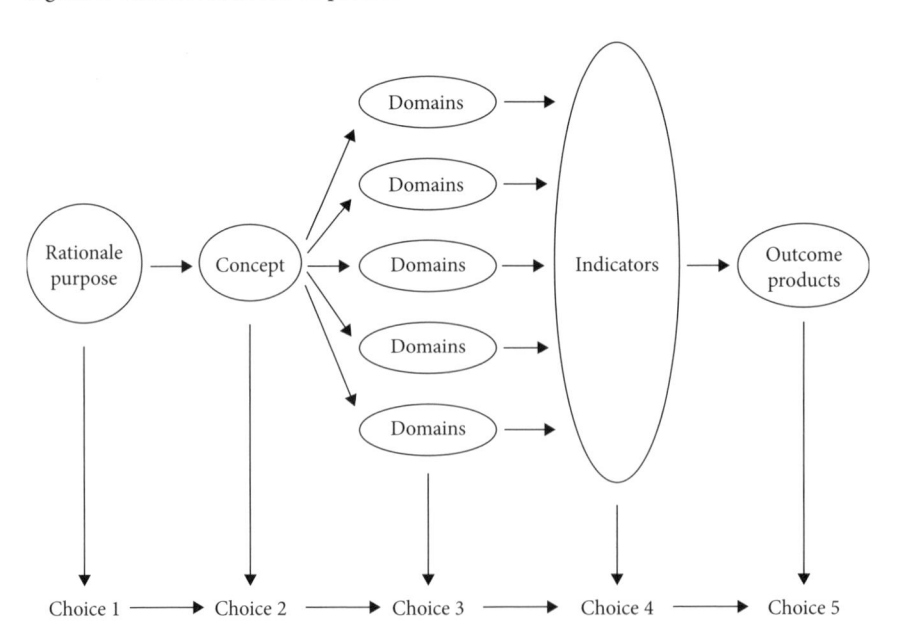

Source: Roelen et al., 2009a.

The generic construction process consists of five separate, consecutive steps that each involve a (set of) choices and decisions (see Roelen et al., 2009a). The first step consists of the establishment of the rationale and purpose of the approach, touching upon the "why" question for the development of a specific approach. The second step focuses on the formulation of the conceptual and theoretical framework of the approach, building a strong foundation for the child poverty approach. The third and fourth steps include the identification of domains and indicators to capture and operationalise the conceptual construct of child poverty. Finally, decisions on the outcome measure and the methodology for its calculation are made in the fifth step.

Decisions made at each step will have implications for the decisions made at the next step and for the combined outcome of the child poverty approach (Roelen et al, 2009a). For example, the purpose of informing the lay public and media about the issue of child poverty might be best served by a measure that captures and summarises child poverty in a single number (see Moore et al., 2007), whilst policy makers might be better served by more detailed information about the underlying domains and characteristics of poverty (see Neubourg et al., 2009). Moreover, the use of different aggregation methods for the estimation of child poverty, decided upon in the last step of the generic construction process, can lead to considerable size and rank differences with respect to child poverty (see Roelen et al., 2009c; Vandivere & McPhee, 2008). In this chapter, we focus specifically on the concept of child poverty and we will use the generic construction process to structure the discussion concerning issues influencing the choice of concept and the implications of this choice.

In the remainder of this chapter, we will firstly focus on a selection of theoretical considerations related to the concept of child poverty. The importance of these considerations within the current debate on child poverty measurement is addressed as well as the reasons for opting for one or the other conceptual notion. Secondly, we illustrate the implications of these theoretical considerations by providing empirical examples. These examples underscore that an understanding of the conceptual construct of child poverty is important beyond the theoretical context and also has practical implications. Finally, we make some concluding remarks on the basis of our analysis.

1. THE CONCEPT OF CHILD POVERTY – THEORETICAL CONSIDERATIONS

In this section, we will discuss in more detail a number of theoretical considerations concerning the concept of child poverty, including the tension between comparative versus contextual definitions of child poverty, the conceptual dichotomy of child well-being versus child well-becoming and the importance of age. Following the logic of the generic construction process, it becomes obvious that specific conceptual choices are a direct result of the approach's rationale and purpose. Heshmati et al. (2008: 189) state: "The main differences are attributed to whether one is interested in monitoring child well-being for the purpose of the evaluation of outcomes and effects of policies or the identification and measurement of impacts of various factors on the outcomes." Roelen et al. (2009a: 250) further emphasise the importance of a proper understanding of a poverty approach's rationale and purpose by stating that "[f]ailing to place the choice for approaches in context of the studies' purpose and rationale, leaves the reader to guess about issues such as multidimensionality, [...] and the unit of analysis". In order to ensure a clear understanding of the considerations in favour of or against specific child poverty concepts discussed in this chapter, we will pay close attention to underlying rationales and purposes. Furthermore, we point towards implications for concurrent steps within the generic construction process, namely the choice of domains, indicators and outcome products, and child poverty outcomes.

1.1. UNIVERSAL VERSUS CONTEXT-SPECIFIC

This first conceptual consideration is one that is not exclusive to child poverty approaches but holds for all poverty approaches and concerns the degree of universality of a poverty approach. To what extent is it desirable for a poverty approach to capture poverty in a multitude of contexts or to reflect one specific context in detail? An answer to this question directly follows from the formulation of the rationale and purpose of the envisaged poverty approach. If the main purpose of a poverty approach is to capture cross-country differences, for example, one requires an approach that is harmonised and comparable across countries. By the same token, if one aims to investigate in-country differences, a poverty approach that captures and reflects a country's societal and cultural context is more desirable.

A review of the body of research on child poverty shows that the majority of studies focus on cross-country comparative analysis (Roelen & Gassmann, 2008). Cross-country studies of child poverty include those by Bradbury and Jäntti (2001), Gordon et al. (2003a, 2003b), Bradshaw et al. (2006), Save the Children (2008) and Richardson et al. (2008). Whilst Gordon et al. (2003a, 2003b) focus on developing countries, Bradbury and Jäntti (2001) study industrialised countries, Bradshaw et al. (2006) focus on countries in the EU and Richardson et al. (2008) on the CEE/CIS countries. Save the Children (2008) takes a global perspective, incorporating countries from around the world. A few examples of studies focused on measuring child poverty in a specific country include the work by Roelen et al. (2009a, 2009b) in Vietnam, Noble, Wright and Cluver (2006) and Barnes et al. (2007) in South Africa and Land, Lamb and Mustillo (2001) in the United States.

Generally, poverty approaches aiming to capture and explain cross-country differences are more restrictive with respect to their design as the construct of poverty has to take the same meaning and form across countries. Moreover, taking a cross-country comparative focus implies that data on the same set of issues has to be available for all countries, the choice of domains and indicators has to present an adequate reflection of child poverty in all countries and that thresholds have to be defined meaningfully for all countries under consideration (see Roelen et al. 2008). Studies aiming to capture and explain in-country variation of child poverty are not bound by these considerations and can rely on country-specific data, reflecting issues that are relevant in the national context. Whilst it might be possible to include issues of cross-country relevance in a context-specific poverty approach, it is not feasible to incorporate issues specific to a single country in a comparative approach. With respect to this conceptual consideration, Thorbecke (2008) refers to conflicts between consistency and specificity criteria. The concept will either focus on a harmonised and more universal construct of poverty or on a context-specific framework with an explicit reflection of societal and cultural characteristics. The choice of domains and indicators, outcome measures and aggregation measures will consequently also be affected (see Vandivere & McPhee, 2008). A discussion of the implications of the consideration of a comparative versus context-specific concept in more empirical terms follows in the next section.

1.2. WELL-BEING VERSUS WELL-BECOMING

The second conceptual consideration discussed in this chapter is specific to child poverty measurement and refers to the dichotomy between the concepts of well-being and well-becoming. There is widespread agreement that poverty during childhood has adverse life-long effects and damages the future development of a child (Brooks-Gunn & Duncan, 1997; Duncan & Brooks-Gunn, 1997; Esping-Andersen & Sarasa, 2002). Duncan and Brooks-Gunn (1997) state that events, environmental conditions and the contexts in which children reside influence the skills and competencies that they acquire. In other words, a denial of child well-being in the present hampers a child's well-becoming in the future. Child poverty approaches grounded in the conceptual construct of well-becoming are thus more focused on issues pertaining to the possibilities for a child to develop and grow up to have an adequate level of well-being in adulthood.

However, the focus on child well-being now is not merely justifiable on the basis of its implications for the future but also because of its importance here and now (Ben-Arieh, 2000; Qvortrup, 1997). Qvortrup (1999) expresses the fear that a sole focus on child well-becoming "[…] justifies any type of life for children, provided the end result – that is, the adult person – exhibits positive values on a set of success criteria". The intrinsic importance of child well-being also follows the concept of children's rights (Ben-Arieh, 2000) as stipulated in the Convention on the Rights of the Child (UNHCHR, 1989). Child poverty approaches built on the concept of well-being are more likely to incorporate issues reflecting children's well-being in the present without taking their potential future influences into account.

The theories of both well-being and well-becoming influence the concept of child poverty and will have their implications for the remaining steps in the generic construction process and child poverty estimates. Poverty approaches adhering to the theory of well-becoming have also been referred to as opportunity-based (Robeyns, 2003) and ex-ante approaches (Thorbecke, 2008). This type of instrumental approach focuses on the capabilities, opportunities or instruments that an individual has at his or her disposal to create favourable outcomes (see Sen, 1999; Wagle, 2002; Robeyns, 2003) and the consequent choice of domains and indicators will focus on reflecting children's opportunities or capabilities to develop and grow, such as education, favourable living conditions and positive influences. As children are highly dependent on their direct environment for the creation of opportunities and favourable conditions (White et al., 2003), the set of indicators is also likely to include a set of contextual indicators. Contextual indicators are those indicators that "[...] pertain to aspects of children's environments that can influence their well-being" (Moore et al., 2007), including the situation of the head of household, family and community. Poverty approaches that are framed along the lines of well-being are considered outcome-based (Robeyns, 2003) or ex-post approaches (Thorbecke, 2008). The approaches capture outcomes with intrinsic value rather than opportunities and focus on the situation as it presents itself to the child at a given point in time. Issues that are relevant within this context focus on the quality of life (see Ben-Arieh, 2000) and might include leisure activities, a safe environment and life satisfaction.

Despite the distinct focus of both conceptual constructs, they are also complementary and can be synergetic. In his seminal work on the basic needs approach, Streeten (1984: 976) had already postulated that "[t]he consumption aspects and the investment aspects of human resource development thus reinforce each other". The main reason for the existence of a wide disparity between the two schools of thought rather than a harmonisation of theories follows from an exclusive focus on either the human or productive aspect (Streeten, 1984). Although the conceptual dichotomy persists, many child poverty approaches do not clearly distinguish between the concepts of well-being versus well-becoming and propose hybrid concepts that combine aspects of both (Ben-Arieh, 2000). The intrinsic value as well as future importance of child poverty and well-being calls for an interest in both present and future childhood (Qvortrup, 1997). The identification of the rationale and purpose will usually not instigate a distinct choice of either one of the approaches given the complementarity of both issues in children's lives. A bias, albeit implicit, towards either the concept of well-being or well-becoming, however, has implications for the consequent development of child poverty approaches and their empirical outcomes.

1.3. AGE MATTERS

The consideration of age and age diversification is another important aspect of child poverty concepts. In other words, are children conceptualised as one homogeneous group or are differences on the basis of age incorporated? Generally, child poverty approaches are developed for children at large, with limited diversification between age groups. Unless the purpose of the child poverty approach clearly stipulates the analysis of a specific age group, a one-size-fits-all type of concept is applied. The comparative child poverty studies by Gordon et al. (2003a, 2003b) and Bradshaw et al. (2006) as well as the country-specific studies by Noble et al. (2006) and Roelen et al. (2009) employ a single concept for children in different stages of childhood. Moore et al. (2008) present a rare example of an age-specific child poverty study with an age-appropriate construct of child poverty by focusing on middle childhood. The limited development and use of age-appropriate child poverty concepts does not mean that there is no age diversification with respect to child poverty approaches. However, the diversification is often not part of the conceptual considerations of the child poverty approach but occurs at later stages of the generic construction process.

Adaptations of the poverty approaches for children in different age groups are the result of practical considerations rather than the formulation of an age-appropriate concept following from a specific rationale or purpose. The availability of data is the primary practical consideration for age diversification and leads to the inclusion of different indicators in child poverty approaches for children in different age groups. Hence, data availability (or lack thereof) can unintentionally divert the final child poverty estimates from their conceptual construct. This kind of diversification by age group does, however, lack a clear theoretical and conceptual foundation, whilst age diversification of child poverty approaches on the basis of conceptual grounds would ensure more solid and transparent measures of child poverty.

Age-diversified concepts of child poverty can be informed by the purpose of an approach to capture a specific age group as well as by developmental theory. An increasing body of research that incorporates developmental theory indicates that the timing of poverty during childhood matters (Duncan & Brooks-Gunn,

1997). Low levels of income and the receipt of welfare in the early childhood period have been found to have a greater impact on school dropout rates than they do in other periods (Duncan & Brooks-Gunn, 1997; Duncan et al., 1998). Moreover, deep and persistent poverty in early childhood is likely to have more perverse effects in comparison to such episodes in later years of life (Duncan & Brooks-Gunn, 2000). Despite the notion that age matters, it has led to only a limited reflection in poverty measures. Developmental theory may offer some support to the identification of different aspects of a similar dimension that are relevant for children in different life stages (see Duncan & Brooks-Gunn, 2000). In terms of parental influence, it is postulated that learning activities at home are important in the early years, parental supervision of homework is important in later elementary and early secondary school years and parental monitoring of friendships is important in adolescent years (Duncan & Brooks-Gunn, 2000). Moore et al. (2008) argue that middle childhood is a stage in children's lives with important development and may in fact be more important for the prediction of future outcomes than early childhood.

On the basis of these premises, a comprehensive concept and a list of indicators are formulated to capture their specific context (Moore et al. 2008). In sum, there are sufficient theoretical grounds to opt for age-appropriate concepts of child poverty rather than a one-size-fits-all concept. Such a conceptual diversification might also be better equipped to respond to pragmatic limitations. In terms of empirical implications, the theoretical discussion indicates that outcomes can diverge due to conceptual as well as pragmatic choices. It is especially those pragmatic choices leading to age diversification within the child poverty approach that might lead to misinterpretation of empirical outcomes. The next section discusses these issues in more detail.

2. THE CONCEPT OF CHILD POVERTY – EMPIRICAL IMPLICATIONS

In this section we build on the theoretical considerations put forward in section 2 and consider the empirical implications in further detail.

2.1. UNIVERSAL VERSUS CONTEXT-SPECIFIC

The empirical implications of the use of a universal or comparative versus context-specific concept of child poverty for child poverty outcomes are illustrated by a case study of Vietnam. Roelen et al. (2008) compared child poverty outcomes for Vietnam on the basis of a global and country-specific approach. Commonalities between both approaches include their multidimensional nature, children's rights and basic needs as part of the conceptual framework and the use of quantitative data for the estimation of child poverty. The approaches differ with respect to their purpose and consequently concept in terms of the consistency versus specificity criteria (see Thorbecke, 2008). The global approach was designed for cross-country comparative purposes and employed in the study by Gordon et al. (2003a, 2003b) to investigate child poverty in the developing world. The country-specific approach was designed for the specific case of Vietnam and intentionally aimed to capture Vietnam-specific cultural and social aspects of child poverty (Roelen et al. 2008, 2009a, 2009b). The impact of this conceptual differentiation on the empirical outcomes is exactly what we wish to illustrate here.

Figure 2. Child poverty rates in Vietnam

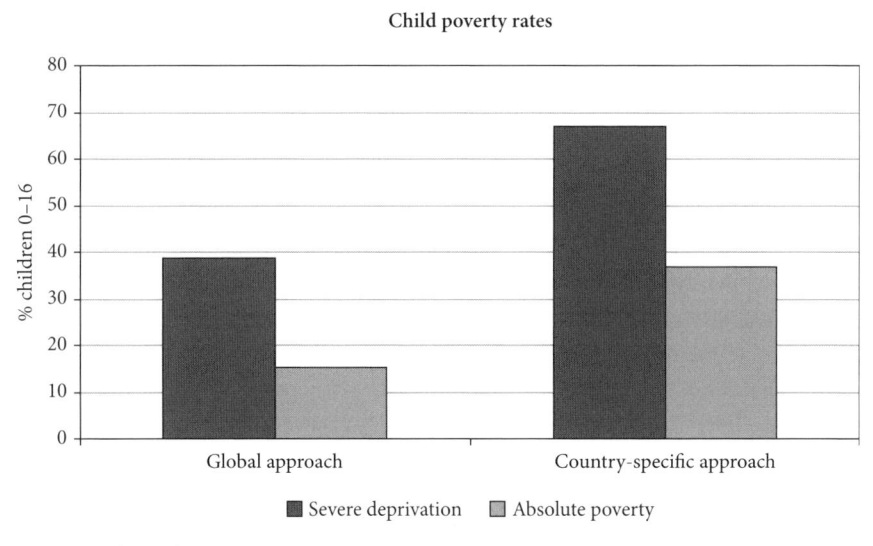

Child poverty rates

[Bar chart showing child poverty rates (% children 0–16). Global approach: Severe deprivation ~39, Absolute poverty ~15. Country-specific approach: Severe deprivation ~67, Absolute poverty ~37. Legend: Severe deprivation, Absolute poverty.]

Source: Roelen et al., 2008.

Figure 2 presents the child poverty incidence rates for Vietnam on the basis of the global and country-specific approach and clearly points towards the existence of size differences. The bar graph represents incidence rates for two different poverty lines, referred to as severe deprivation and absolute poverty.[1] The country-specific approach clearly identifies a larger proportion of the child population in Vietnam as being poor than the global approach. Whilst 39 percent of all children aged 0–16 are considered severely deprived and 15 percent are considered absolutely poor according to the global approach, the proportions are respectively 67 percent and 37 percent according to the country-specific

---

[1] Please refer to Roelen et al. (2008) for a detailed description of the approaches' methodologies.

approach. Child poverty outcomes on the basis of both approaches do not, however, point to significant group differences (Roelen et al., 2008). Hence, the country-specific approach captures more children in Vietnam than does the global approach but it does not capture a different group of children.

The conceptual framework of the child poverty approaches is important for explaining and understanding the differences in the estimated magnitude of child poverty. Firstly, the universality aspect within the concept of the global approach instigated the choice of domains and indicators that were applicable and relevant for all countries, whilst the country-specific approach was not constrained by this criterion. As a result, the domains of labour, leisure and social protection were part of the country-specific but not of the global approach (Roelen et al., 2008). Given that particularly the domain of leisure reported high rates of deprivation, the child poverty rates are consequently more pronounced in the case of the country-specific approach. Secondly, the large diversity between developing countries, for which the global approach was developed, and the high degree of deprivation in some countries caused the global approach to err on the side of caution with respect to the determination of thresholds (Gordon et al., 2003; Roelen et al., 2008).

The example of a specific indicator in the dimension of education illustrates this point: "While the country-specific approach considers net enrollment as an appropriate determinant for education vulnerability, the global approach considers whether a child has ever attended school or not." (Roelen et al., 2008: 15). The large difference between indicator thresholds leads to large discrepancies between the estimated sizes of child poverty. The notion that child poverty approaches aiming to analyse diverse groups of children in a comparative manner lead to different findings from approaches focusing on context-specific issues that children face was also postulated by Vandivere and McPhee (2008). The large degree to which the differences between concepts lead to diverging outcomes makes the specificity versus consistency notion relevant beyond the realm of theory. If child poverty estimates are designed to capture the cross-country comparative situation, they should be interpreted in that context. By the same token, estimates based on country-specific approaches should not be employed to evaluate and interpret the situation relative to other countries (see also Moore et al., 2007).

2.2. WELL-BEING VERSUS WELL-BECOMING

The empirical implications of the conceptual tension between child well-becoming and child well-being are partly reflected in the choice of domains and indicators it inspires. As discussed previously, many child poverty approaches do not explicitly adopt either of these conceptual constructs but rather employ an albeit implicit hybrid form. As such, we cannot compare the outcomes of a child well-becoming approach with those of a child well-being approach. Nevertheless, we can consider outcomes at indicator level, comparing deprivation rates for an indicator inspired by the concept of well-being and child life satisfaction and an indicator reflecting an instrument for future well-being and thus the concept of well-becoming.

Figure 3 presents Vietnamese deprivation rates for indicators in the domains of education and leisure disaggregated by region. Education is an area that is often associated with well-becoming of children and with the creation of capabilities to have a good future life (see Brooks-Gunn & Duncan, 1997; Duncan et al., 2000). Leisure can be considered an area that is important for a child's development but also as an aspect of childhood with great intrinsic importance; it constitutes a factor for the well-being and quality of life of the child in the here and now (see Ben-Arieh, 2000). The education indicator refers to the percentage of children of school age that are not net-enrolled. The leisure indicator reflects the percentage of children below the age of five that do not have any toys available (Roelen et al., 2009c).[2] The most notable observation from the deprivation incidence rates in Figure 3 is the rank differences between regions. Whilst some regions fare relatively well with respect to one indicator, their performance with respect to the other indicator is relatively poor.

Figure 3. Education and leisure deprivation in Vietnam

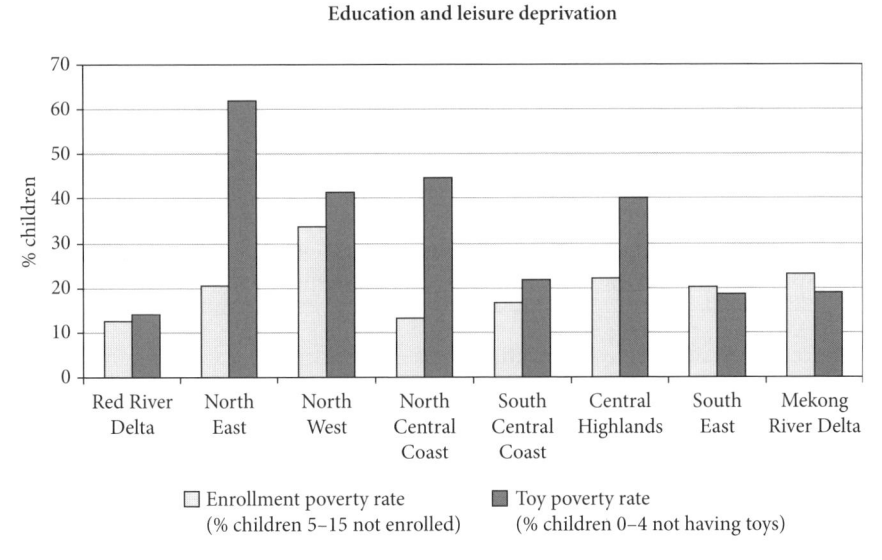

Source: Roelen et al., 2009c.

---

[2] This specific example focuses on the illustration of differences in outcomes between indicators grounded in the well-becoming versus well-being concept; the difficulties and need for caution in interpreting the results of indicators referring to children in different age groups are addressed in following section.

The analysis of these indicators shows that an assessment of child deprivation based on an approach biased towards the concepts of either child well-being or child well-becoming might lead to divergent pictures. In this specific example, the inclusion of the child well-being perspective will lead to the issue of child deprivation being considered to be greater in magnitude and to be most pronounced in the North East and North Central Coast regions. A focus on child well-becoming will provide a more favourable overall picture of child deprivation and point to the North West and Mekong River Delta as regions that require a most urgent improvement of their situation. Regional rankings might pose an incentive for poor-performing regions to improve their child poverty situation and as such inform policy decisions and budget allocations (Vandivere & McPhee, 2008). In light of this argument, this example indicates the importance of understanding the concepts underlying child poverty estimates. Policy makers need to be aware of and understand the actual issues underlying child poverty outcomes to ensure that they are responding to the problems that they wish to address in specific regions.

2.3. AGE MATTERS

The discussion pertaining to the empirical implications of the age consideration will focus on the diversification of child poverty outcomes by age resulting from pragmatic rather than conceptual considerations. To our knowledge, there are no studies that have attempted to compare child poverty outcomes for children in different age brackets on the basis of age-diversified concepts of child poverty. As such, we cannot compare empirical outcomes for an age versus non-age diversified approach and look into possible size, rank or group differences. Instead, we consider child poverty estimates that are diversified by age due to practical and pragmatic considerations and to what extent they may be susceptible to misinterpretation and misunderstanding.

Figure 4 presents severe deprivation rates for children in the developing world in different dimensions of basic needs. The figure is taken from the study by Gordon et al. (2003a, 2003b), which compares the status of child poverty in 46 developing countries. The study implicitly treats children within the entire age spectrum between 0–18 as a homogeneous group and does not diversify its concept of poverty for children in different age brackets. Nevertheless, the final list of indicators employed for the measurement of child poverty is age-specific with estimates referring to children in different age brackets. This is largely due to data availability and the fact that specific information within the respective surveys[3] is not available for children of all ages. In other words, many surveys have acted upon the notion that different issues are important for children of different ages and have designed their questionnaires accordingly. Child poverty approaches and studies, however, tend to respond to the availability of data on specific indicators for specific age groups in an ad hoc and pragmatic manner instead of changing its structural foundation. Such a response results in the presentation of poverty outcomes as illustrated in Figure 4.

Figure 4. Severe deprivation rates for children

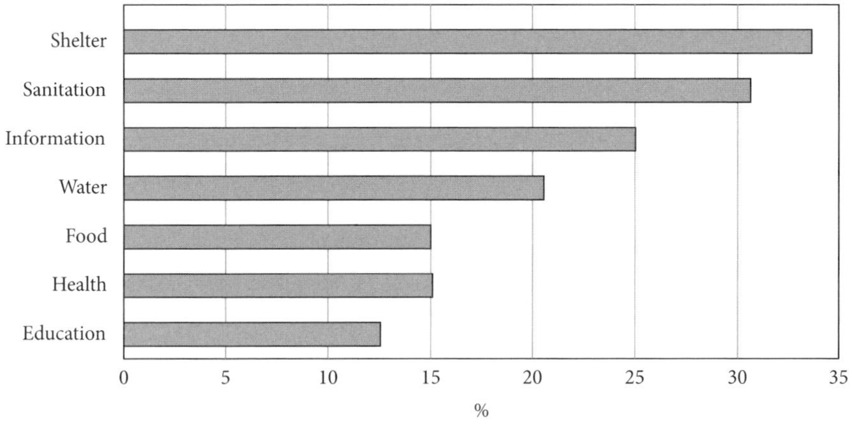

```
Shelter      ████████████████████████████████████  ~34
Sanitation   ███████████████████████████████       ~31
Information  █████████████████████████             ~25
Water        █████████████████████                 ~21
Food         ███████████████                       ~15
Health       ███████████████                       ~15
Education    █████████████                         ~13
             0    5    10   15   20   25   30   35
                              %
```

Source: Gordon et al., 2003.

The domain deprivation rates represent the proportion of children that are severely deprived with respect to these domains. Estimates indicate that approximately 34 percent of children are deprived with respect to shelter, 15 percent are food deprived and 13 percent are educationally deprived. This straightforward interpretation suggests that the estimates capture all children in all age brackets, without any kind of differentiation between age groups. However, a closer look at the underlying indicators and their reference populations reveals that only certain domains account for the total child population whilst other domains pertain to children in a limited age group. The domain of shelter captures the total child population, the food domain refers to children less than five years of age and the education domain pertains to children of school age. A comparison of domain deprivation estimates as presented in Figure 4 should thus be undertaken with caution as they do not

---

[3] The study by Gordon et al. (2003a, 2003b) is based on Multiple Indicator Cluster Surveys (MICS) and Demographic and Health and Demographic Surveys (DHS) for the countries under consideration.

necessarily reflect issues for all children. One might be tempted to draw conclusions about the situation with respect to nutrition for older children or pre-school education on the basis of these outcomes, although this is not actually reflected.

The fact that this age diversification is not reflected in the concept of child poverty and not clearly highlighted in the presentation of results makes the estimates susceptible to misinterpretation. A more explicit and transparent presentation of results by researchers and a more in-depth and closer look into the underlying framework of the child poverty approach by its users can avoid such misinterpretations and ensure appropriate understanding and use of child poverty estimates.

3. CONCLUSION

At the outset of this chapter, it was stated that a clear and transparent insight into and understanding of the concept underlying child poverty approaches is not merely interesting from a philosophical or scientific point of view but also from a policy and user perspective. The development of child poverty approaches is a normative process, hinging on value judgments and theoretical considerations and leading to different interpretations of reality. As a result, child poverty outcomes may diverge depending on the approach taken. A lack of transparency or understanding of the various choices made during the development process of child poverty approaches, including those on definitions, concepts and methodology, can cause misinterpretation of those outcomes and result in inappropriate and inadequate policy responses. This chapter specifically focused on the conceptual choices and considerations and their empirical consequences. The strong link between concept and outcomes is exactly what makes knowledge about this concept so crucial in understanding and correctly using child poverty outcomes and the various examples in this chapter have served as an illustration of this.

The theoretical considerations with respect to the concept of child poverty, the factors inspiring conceptual choices as well as the implications for the remaining steps in formulating a child poverty approach are naturally of interest and value to those developing child poverty approaches. However, this discussion is equally important for those using and working with child poverty estimates, including policy makers and practitioners. In order to respond adequately to the issue of child poverty and its underlying factors, one needs to have a proper understanding of what the term "child poverty" and its estimates actually represent. Do estimates reflect the situation on the basis of internationally comparable standards or country-specific criteria? Do figures capture issues pertaining to children's well-being here and now or focus on instrumental aspects that account for well-becoming in the future? Do rates account for a specific age group of children and needs and requirements specific to the stage of childhood or consider children of all ages as one homogeneous group? Having an answer to these questions and others concerning the actual meaning underlying the term "child poverty" is crucial for the establishment of a common ground for interpreting and discussing the issue of child poverty, formulating responses to the issue, designing appropriate policies and monitoring them effectively.

Focusing the discussion on a limited range of conceptual issues implies that we leave a number of issues out of consideration, despite their importance. One of the conceptual considerations that has been and remains a subject of extensive debate is the disparity between monetary versus multidimensional poverty measurement. One might argue that the tension between monetary and multidimensional poverty measurement is absorbed by the discussion on well-becoming versus well-being; especially with respect to children, income can be considered a means to an end rather than an end in itself. However, we did not want to reduce the longstanding and complex debate on monetary versus multidimensional child poverty measurement to the well-becoming versus well-being dichotomy. Whilst one might argue that monetary approaches exclusively capture children's well-becoming, the reverse argument might not hold with respect to multidimensional child poverty measures as they could be based on the concepts of both well-becoming and well-being or a hybrid form.

This chapter has also largely disregarded other aspects identified by the generic process that potentially lead to diverging child poverty outcomes. The importance of data availability (or lack thereof) and how it can unintentionally divert the final child poverty estimates from their conceptual construct has been considered in the discussion on age diversification. Nevertheless, given the significance of this issue and its potential impact on the development of child poverty approaches and consequent outcomes, it requires more elaborate consideration. Furthermore, the choice of outcome products and methodologies for their calculation potentially leads to diverging pictures of child poverty. Future research should focus on these and other aspects and address their impact on the development, measurement and use of child poverty approaches.

REFERENCES

ALKIRE, S. (2008), Choosing dimensions: the capability approach and multidimensional poverty, in: N. KAKWANI & J. SILBER (eds.), *The many dimensions of poverty*. New York: Palgrave-Macmillan.

BARNES, H., WRIGHT, G., NOBLE, M. et al. (2007), *The South African index of multiple deprivation for children: census 2001*. Cape Town: HSRC Press

BEN-ARIEH, A. (2000), Beyond welfare: measuring and monitoring the state of children – new trends and domains, *Social Indicators Research*, (52): 3, 235–257.

BRADBURY, B. & JANTTI, M. (2001), Child poverty across industrialised world: evidence from the Luxembourg income study, in: K. VLEMINCKX & T. SMEEDING (eds.), *Child well-being: child poverty and child policy in modern nations: what do we know?* Bristol: The Policy Press.

BRADSHAW, J., HOELSCHER, P. & RICHARDSON, D. (2006), An index of child well-being in the European Union, *Social Indicators Research*, (80): 1, 133–177.

BROOKS-GUNN, J. & DUNCAN, G. (1997), The effects of poverty on children, *The Future of Children*, (7): 2, 55–71.

DUNCAN, G. & BROOKS-GUNN, J. (1997), *Consequences of growing up poor*. New York: Russell Sage Foundation.

DUNCAN, G. & BROOKS-GUNN, J. (2000), Family poverty, welfare reform, and child development, *Child Development*, (71): 1, 188–196.

DUNCAN, G., YEUNG, J.W., BROOKS-GUNN, J. et al. (1998), How much does childhood poverty affect the life chances of children? *American Sociological Review*, (63): 3, 406–423.

ESPING-ANDERSEN, G. & SARASA, S. (2002), The generational conflict reconsidered, *Journal of European Social Policy*, (12): 1, 5–21.

GORDON, D., NANDY, S., PANTAZIS, C. et al. (2003a), *Child poverty in the developing world*. Bristol: Policy Press.

GORDON, D., NANDY, S., PANTAZIS, C. et al. (2003b), *The distribution of child poverty in the developing world*. Bristol, UK: Centre for International Poverty Research.

HAVEMAN, R. & WOLFE, B. (1995), The determinants of children's attainments: a review of methods and findings, *Journal of Economic Literature*, (33): 4, 1829–1878.

HESHMATI, A., TAUSCH, A. & BAJALAN, C. (2008), Measurement and analysis of child well-being in middle and high income countries, *The European Journal of Comparative Economics*, (5): 2, 227–286.

LADERCHI, C.R. (1997), Poverty and its many dimensions: The role of income as an indicator, *Oxford Development Studies*, (25): 3, 345–360.

LADERCHI, C.R., SAITH, R. & STEWART, F. (2003), Does it matter that we do not agree on the definition of poverty? a comparison of four approaches, *Oxford Development Studies*, (31): 3, 243–274.

LAND, K., LAMB, V. & MUSTILLO, S.K. (2001), Child and youth well-being in the United States, 1975–1998: some findings from a new index, *Social Indicators Research*, (56): 3, 241–320.

MOORE, K.A., VANDIVERE, S., ATIENZA, A. et al. (2008), Developing a monitoring system for indicators in middle childhood: identifying measures, *Child Indicators Research*, (1): 2, 129–155.

MOORE, K.A., VANDIVERE, S., LIPPMAN, L. et al. (2007), An index of the condition of children: the ideal and a less-than ideal US example, *Social Indicators Research*, (84): 3, 291–331.

NEUBOURG, C. D., ROELEN, K. & GASSMANN, F. (2009), Making poverty analyses richer – multidimensional poverty research for policy design, in: K. DE BOYSER, C. DEWILDE, D. DIERCKX & J. FRIEDRICHS (eds.), *Between the social and the spatial – exploring multiple dimensions of poverty and social exclusion*. Farnham: Ashgate Publishing, pp. 35–56.

NOBLE, M., WRIGHT, G. & CLUVER, L. (2006), Developing a child-focused and multidimensional model of child poverty for South Africa, *Journal of Children and Poverty*, (12): 1, 39–53.

QVORTRUP, J. (1997), Indicators of childhood and intergenerational dimension, in: A. BEN ARIEH & H. WINTERSBERGER (eds.), *Monitoring and measuring the state of children – beyond survival*. Vienna: European Centre for Social Welfare Policy and Research, pp. 101–112.

QVORTRUP, J. (1999), The meaning of child's standard of living, in: A. BOWERS ANDREWS & N. HEVENER KAUFMAN (eds.), *Implementing the UN Convention on the Rights of the Child: a standard of living adequate for development*. Westport: Praeger Publishers.

RAVALLION, M. (1994), *Poverty comparisons*. London: Harwood Academic Publishers.

REDMOND, G. (2008), Child poverty and child rights: edging towards a definition, *Journal of Children and Poverty*, (14): 1, 63–82.

RICHARDSON, D., HOELSCHER, P. & BRADSHAW, J. (2008), Child well-being in the CEE and the CIS, *Child Indicators Research*, (1): 3, 211–250.

ROBEYNS, I. (2003), *The capability approach: an interdisciplinary introduction*. Amsterdam: University of Amsterdam.

ROELEN, K. & GASSMANN, F. (2008), *Measuring child poverty and well-being: a literature review*. MGSOG Working Paper 2008/WP001. Maastricht: Maastricht Graduate School of Governance.

ROELEN, K., GASSMANN, F. & NEUBOURG, C.D. (2008), *A global measurement approach versus a country-specific measurement approach: do they draw the same picture of child poverty? The case of Vietnam*. MGSOG Working Paper 2008/WP005. Maastricht: Maastricht Graduate School of Governance.

ROELEN, K., GASSMANN, F. & NEUBOURG, C.D. (2009a), The importance of choice and definition for the measurement of child poverty – the case of Vietnam. *Child Indicators Research*, (2): 3, 245–263.

ROELEN, K., GASSMANN, F. & NEUBOURG, C.D. (2009b), Child poverty in Vietnam – providing insights using a country-specific and multidimensional model. *Social Indicators Research*, DOI: 10.1007/s11205-009-9522-x.

ROELEN, K., GASSMANN, F. & NEUBOURG, C.D. (2009c), *Child poverty in Vietnam – providing insights using a country-specific and multidimensional model*. MGSOG Working Paper 2009/WP001. Maastricht: Maastricht Graduate School of Governance.

SAUNDERS, P. (2004), *Towards a credible poverty framework: from income poverty to deprivation*. SPRC Discussion Paper No. 131. Sydney: SPRC.

SAVE THE CHILDREN (2008), *The Child Development Index*. London: Save the Children UK.
SEN, A. (1999), *Development as freedom*. Oxford: Oxford University Press.
STREETEN, P. (1984), Basic needs: some unsettled questions, *World Development* (12): 9, 973–978.
THORBECKE, E. (2008), Multidimensional poverty: conceptual and measurement issues, in: N. KAKWANI & J. SILBER (eds.), *The many dimensions of poverty*. New York: Palgrave Macmillan.
UNHCHR (1989), *Convention on the Rights of the Child* (General Assembly resolution 44/25). Geneva, UNHCHR.
VANDIVERE, S. & MCPHEE, C. (2008), Methods for tabulating indices of child well-being and context: an illustration and comparison of performance in 13 American states, *Child Indicators Research*, (1): 3, 251–290.
VRANKEN, J. (2010), 'Poverty, like beauty, lies in the eye of the beholder'? in: W. VANDENHOLE, J. VRANKEN & K. DE BOYSER (eds.), *Why Care? Children's rights and child poverty*. Antwerp: Intersentia.
WAGLE, U. (2002), Rethinking poverty: definition and measurement, *International Social Science Journal*, (54): 171, 155–165.
WHITE, H., LEAVY, J. & MASTERS, A. (2003), Comparative perspectives on child poverty: a review of poverty measures, *Journal of Human Development*, (4): 3, 379–396.

8. ESCAPING POVERTY WITH YOUR CHILDREN: THE ROLE OF LABOUR MARKET ACTIVATION, EDUCATION, AND SOCIAL CAPITAL INVESTMENTS

Ingrid SCHOCKAERT and Ides NICAISE

INTRODUCTION

We start from the presumption that to a large extent children's life chances depend on the social inclusion of their parents or other guardians. Parents, as adults, have to function in the broader societal context, which may or may not be favourable for them. Factors affecting general poverty will therefore also affect the living conditions of children, but not necessarily to the same extent. In this chapter we will analyse differences in poverty dynamics between adults and children, and look at the effect on child poverty of factors that are known to influence general poverty such as employment, educational attainment and social participation.

Acknowledging the relationship between children's and their household's poverty, the Belgian Resource Centre for the Fight against Poverty claimed that tackling child poverty should be integrated within an overall strategy for combating poverty (Steunpunt voor bestrijding van de armoede, bestaansonzekerheid en sociale uitsluiting, 2008). However, following the former reasoning about the differential effect of social conditions on households with and without children, we can assume that not all measures have the same effectiveness for child poverty relief. In this paper, we will demonstrate the potential impact of general policy measures such as labour market activation,[1] prevention of school dropout and social capital enhancement on the dynamics of child poverty.

Although since the pioneering work of Bane and Ellwood (1986) the importance of a dynamic perspective on poverty has gradually become more accepted, in

---

[1] In operational terms, we will define activation as temporary subsidised employment.

Belgium its development is still rather uncommon. This is even more the case with respect to child poverty (Steenssens et al., 2008). To the best of our knowledge only Dewilde and Levecque (2003) have looked at the difference in the length of poverty spells between households with and without children.

We believe that further development of a dynamic approach to child poverty in Belgium may offer a useful complementary perspective to the cross-section picture, particularly since factors triggering poverty may well differ from those determining exit from poverty. In addition, by differentiating between persistent, temporary and recurrent poverty the consequences of poverty for children's future life chances can be assessed more accurately, since an isolated short spell is probably in most cases less damaging than a longer period of deprivation or recurrent poverty (Bradbury et al., 2000). In this paper, we will focus on the length of poverty periods and the causes of poverty persistence.

Our methodology is an extension of the ex-post micro-simulation model developed by De Blander and Nicaise (2009). This model offers a method for the dynamic analysis of adult poverty and the simulation of the potential impact of policy measures. The transformation of the poverty risks of adults into poverty risks of their children allows its use for analysing child poverty and evaluating the impact of general policies on its evolution.

The paper is organised as follows. First, we present the theoretical motivation of our analysis. Second, we explain our methodology. Then we present results on the probability of exit from poverty (outflow rates) of adults and children and on the potential impact of activation, prevention of school dropping out and enhancing social capital. Finally, theory and results are brought together in some policy recommendations.

1. THEORETICAL PERSPECTIVE

According to Beck (1992), recent societal changes have led to what he called a "risk society" characterised by intense mobility – upward and downward – induced by processes of globalisation, increased flexibility and technological transformation. Moreover, the life courses of individuals have been "destandardised", making events such as marriage and labour force participation, but also experiences regarding poverty, more dependent on individual decisions and less on social background (Beck, 1992; Myles, 1993; Dewilde, 2003). In Leisering and Leibfried's (1999) rather extreme vision, these societal changes have converted poverty into a merely temporal and "democratised" phenomenon: it is simply an event that can happen to anyone.

We, to the contrary, are not convinced that the increase in mobility and individualisation is necessarily synonymous with a decrease in the role of social differentiation. According to Dieleman (2000), individualisation implies a new standard norm that demands that individuals choose, plan and be responsible for their own life. However, not everyone is equally equipped "freely" to determine its course. Control – over potential choices *and* their outcomes – depends heavily on the resources available to each individual. These resources have to be interpreted in the broad sense, as in Venturini's (2008) concept of "endowment", referring to the "material and nonmaterial resources transmitted by the environment (…), which the individual has accumulated (…) in the past (…)." In a modern 'mobile' society, inequality does not imply a stable division between groups of "haves" and "have-nots", but is a dynamic phenomenon operating through unequal capital accumulation to determine future life chances. Education and social capital are considered to be important resources.

The impact of education on income has been widely confirmed (Blundell et al., 2005) and this is also the case for Belgium (Nicaise, 1998, Groenez et al., 2010). Moreover, the role of education has grown over the years (OECD, 2005). However, Machin and McNally (2006) noted that despite the many studies about the returns to education, there is not much evidence about the returns for specific disadvantaged subgroups, such as the poor. The few studies they analysed showed that the returns to an additional year of schooling were higher at the bottom of the educational and income ladder than for the population as a whole, but that this was also dependent on the qualification and type of training received. In addition, McIntosh (2004) showed that even though additional vocational training did not increase earnings, it had a large impact on the probability of gaining employment.

The social capital concept refers to social relations of trust and reciprocity, which can increase people's wellbeing and facilitate human and material capital accumulation (Bourdieu, 1993; Putnam, 1993; Coleman, 1988). A distinction can be made between "bonding", "bridging" and "linking" social capital (World Bank, 1998). The former refers to bonds between family members, friends and neighbours, involving a strong sense of trust and reciprocity. They can serve as a "safety net", but also provide psychological and emotional support that makes people more resilient. In contrast, some authors point out that these bonds can lead to social isolation from the rest of society or downward social pressure impeding mobility more than facilitating it (Portes, 1998). "Bridging" social capital implies weaker ties originating in associations. They serve as "bridges" that fill the structural blanks between several networks of strong ties. These contacts seem to be springboards for social mobility, for example in finding a job. "Linking" social capital consists of links between groups of different social classes or power holders and seems to be fairly rare. In Belgium, the notion was taken up in poverty research during the 1990s (Steenssens et al., 1996) and developed into an important concept (Hooghe, 1999; Peeters & Vranken, 2000; Thys, 2001; Thys et al., 2004; De Boyser & Levecque, 2007; Groenez et al., 2010).

While education, training and social capital formation amplify an individual's opportunity set and thus future life chances, activation policies aim at facilitating and encouraging certain types of behaviour which are considered adequate for social inclusion – that is, to be active in the labour market – by conditioning support to the individual's willingness to take up this economic responsibility. It is in this aspect of the "active welfare state" that society's individualisation tendency is most clearly adopted. The results of such an activation policy, however, are not completely unambiguous. Some authors find a positive but fading impact on the mobility out of poverty or social welfare (Autor & Houseman, 2005; Cockx & Ridder, 2001; De Blander & Nicaise, 2005); others discover a more durable effect (Ayala & Rodriguez, 2006; Lorentzen & Dahl, 2005). The literature also points out that not every individual is as easily activated. Age, educational attainment and prior work experience (Dahl & Lorentzen, 2005) as well as ethnic characteristics (Alaya & Rodriguez, 2007) influence the results.

Interesting in this regard is the analysis of Dahl and Lorentzen (2007) concerning "employability": the more characteristics an individual shows that make him suitable for the labour market, the higher the long-term reduction in poverty risk following activation. Congruently, Hermans (2005) identified the difficulties for activation institutions (the Public Centres for Social Welfare in particular) in matching labour supply and demand, leading to success only for the "best" clients. The debatable success of activation can therefore, according to Dahl (2007), be attributed to the implicit presumption that the person activated possesses the *ability* to convert work into social inclusion, whereas poverty implies precisely the deprivation of these basic capacities. Steenssens et al. (2009) differentiate in this regard "restrictive" – problematisation of the subject – and "reflexive" activation – aiming at the empowerment of the socially excluded. In other words, it seems that activation is most effective when it is accompanied by capacity-building and enhancement of people's resources.

Although the necessity for integrating the fight against child poverty into general poverty relief policy is well recognised (Steunpunt voor bestrijding van de armoede, bestaansonzekerheid en sociale uitsluiting, 2008), little is known either about the differential impact of factors such as education and social capital on adults and children or about the influence of general policy measures on children. The assessment of this difference can, in our opinion, contribute to understanding the causes of child poverty and the potential impact of policy measures.

For parents, all decisions during the life course are mediated by the necessity to balance their care and financial responsibilities, often prioritising the latter. This complex organisation of daily life has an influence on how resources such as education and social capital can be used and converted into social inclusion. Education not only offers access to better paid jobs – with earnings sufficient to compensate child care costs – but also increases the window of opportunity to find work compatible with care. In addition, education enhances individuals' decision making and planning skills independently to find a way to reconcile work and child care. The same argument applies to social capital, which offers a network structure facilitating job seeking and childcare support. As an empowering resource it may help in finding solutions and offer much needed psychological and pedagogical support.

Due to the loss of income, unemployment will make households with children more vulnerable to poverty than households without children. This is shown by the extreme poverty risk for households with children and low or non-existent labour market participation. In 2008, 74.7% of them were poor, compared to 32.2% when no children were present (FPS Economy, SMEs, Middle Classes and Energy, 2010). So at first sight, activation measures have a large potential impact on child poverty. Nevertheless, if parents need to balance between their care and financial responsibilities, activation policies' exclusive emphasis on the latter may well have counterproductive results. Low-income parents lack access to childcare facilities, making job-seeking, labour force participation and job retention more difficult. In addition, even with work it might not be so evident for people with (many) children to convert work into a lower risk of poverty – the income increase related to labour market participation should be high enough to cover additional expenses related to childcare outsourcing. This is not guaranteed if activation is not accompanied by supplementary investment in human capital and capacity building.

2. DATA

Our analysis makes use of eight waves of the Belgian Household Panel Survey, 1993 to 2000. Individuals between the age of 16 and 65 were selected if they had ended full-time education. This selection ensures the homogeneity of the variables work and educational level for the whole sample.

Poverty is measured through total net household income. Yearly income is converted to monthly income, based on the individual's monthly work status and information on social transfers during the year (Nicaise et al., 2004). Income is then grouped into a trichotomous variable that differentiates between people with a household income (a) below or equal to the social assistance level, (b) between the social assistance threshold and the "at-risk-of-poverty" threshold (60% of the median equivalent income) and (c) at or above the latter level. These categories will be referred to as "at risk of poverty", "underprotected" and "non-poor" respectively. For each individual we have between 12 and 96 observations, depending on the moment of entry into the panel and possible panel exit.

Educational status is operationalised through a discrete variable with four levels indicating the highest degree obtained: no or only primary education, lower secondary education, upper secondary education and higher education. Since our sample only contains individuals who have left school and since we did not take into account additional schooling after entering the labour market, this is a time independent variable.

Work is a discrete variable representing the respondent's work status for each month of observation. Partners of self-employed workers who claimed that they were "helping" their spouses were considered to be working.

Social capital is measured through a series of indicators of social contacts and participation in associational life: frequency of contact with neighbours ("Neighbours") and with friends ("Friends"), voluntary work ("Volunteer"), association membership ("Member Org"), weekly participation in organised sports activities ("Weekly Sport") and weekly participation in organised creative activities ("Weekly Creative"). Information is available on a yearly basis from 1994 onwards. These indicators were all submitted to a multiple correspondence analysis withholding two dimensions presented as the X and Y axes in Figure 1 (Annex 1). About 37% of all inertia was represented.

Contacts with friends and neighbours show high scores on the Y-axis. Moreover, the frequency of contacts shows a continuum: few or no contacts with friends and neighbours (Neighbours.1 and Friends.1) have high positive scores, while high frequency contacts (Neighbours.3, Neighbours.4, Friends.3 and Friends.4) have negative scores. The categories Neighbours.2 and Friends.2 fall in between (Neighbours.5 indicates missing values). On the X-axis we find the categories of variables representing a more formal participation in social life. Negative scores indicate absence of participation, while positive scores point to some degree of integration into associative life. Note that all categories are very close to one of the axes. In other words, the analysis confirms that bonding and bridging social capital are different concepts. They will be treated separately in the rest of our analysis.

Our analysis of social capital is mostly consistent with the results found by Groenez et al. (2010). In spite of differences in method (factor analysis), data

(PSBH 1998) and additional variables included (indicators of social and cultural activities), they also found a clear difference between the dimensions measuring formal and informal social network participation. The main difference consists of the lack of significant factor loadings on either dimension of the variables measuring weekly sports and creative activity.

3. METHOD

Building on the methodology proposed by De Blander and Nicaise (2009), our research strategy consists of an estimation model and a simulation model, for both adults and children.

3.1. THE MODEL ESTIMATION

We built a series of Markov models[2] measuring as accurately as possible the impact of work, education and social capital on the dynamics of poverty. We made use of a multi-stage selection model, according to the method of generalised residuals (Heckman, 1979; Dubin & McFadden, 1984), taking into account potential attrition[3] and the endogeneity of the initial poverty status and work, education and social capital (called "key variables" in the rest of the paper). As shown in Fig. 2, we first estimated the probability of attrition, then of the respondents being at work (initial employment status as well as subsequent transition rates between employment and non-employment), their educational or social capital level[4], and finally their risk of poverty (initial status and transition rates), each time correcting for the potential correlation between the error terms of the current and previous models. At each step, a series of control variables known to influence the dependent variable is also included. For the estimation of the initial poverty state, in addition to the key variables, we introduced as control variables the respondent's age, gender, household composition (size, number of children, single parenthood, marital status), health status, migrant status (distinguishing between European and non European) and geographical context (Flanders, Wallonia or Brussels region, and urban versus rural context). For the dynamic estimation of poverty, in addition to these

---

[2] A Markov model predicts the state of an individual at point t, dependent on his state at point t-1 (and not additionally on his state at all previous points). In our case, these states are poverty states "non-poor", "at risk of poverty" or "underprotected".

[3] Attrition refers to panel exit due to refusal to cooperate further or to loss of contact. Since attrition is related to poverty and other factors of social stratification, the risk of biased results in panel studies must be corrected for.

[4] These key variables are introduced separately. This means that we have a selection model for work, education and social capital.

control variables, we controlled for life events such as job losses, births, separations, formations of new relationships and changes in health status.

Figure 2. Schematic representation of the multistage selection model for the estimation of dynamic poverty risks

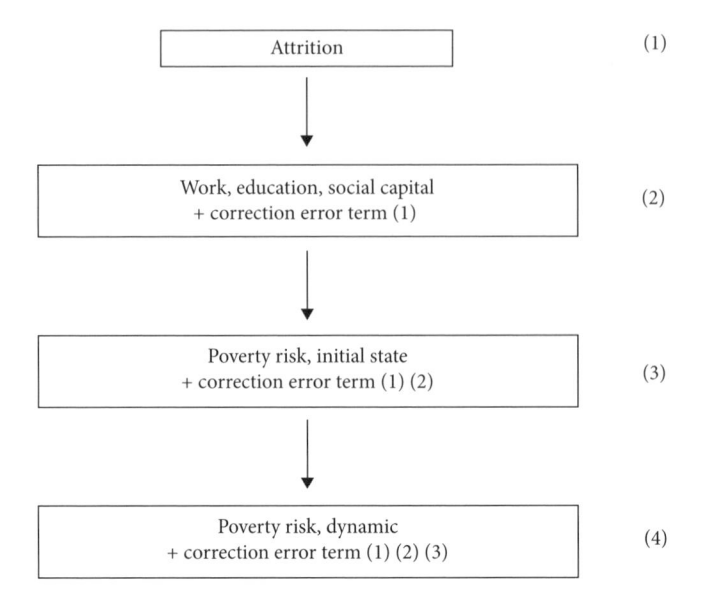

```
         ┌─────────────────────────────┐
         │          Attrition          │        (1)
         └─────────────┬───────────────┘
                       ▼
    ┌──────────────────────────────────────┐
    │   Work, education, social capital    │   (2)
    │      + correction error term (1)     │
    └──────────────────┬───────────────────┘
                       ▼
    ┌──────────────────────────────────────┐
    │       Poverty risk, initial state    │   (3)
    │    + correction error term (1) (2)   │
    └──────────────────┬───────────────────┘
                       ▼
    ┌──────────────────────────────────────┐
    │         Poverty risk, dynamic        │   (4)
    │  + correction error term (1) (2) (3) │
    └──────────────────────────────────────┘
```

For each consecutive estimation, we used a pooled probit, a pooled ordered probit or a least squares regression, depending on the nature of the dependent variable. For a formal expression of this model we refer to De Blander and Nicaise (2009). The estimation results for poverty are presented in table 1 of annex 1. We observe that work, education and bridging social capital have a positive effect on income. The influence is largest if individuals are underprotected, and smallest if they are at risk of poverty. Note that the variable bonding social capital only has a small significant impact on income for the group of individuals at risk of poverty. Therefore, this variable will not be used in the rest of our analysis.

3.2. THE SIMULATION MODEL

Based on these estimations, we can calculate a transition matrix indicating for each individual and for each month of observation the chance of being non-poor, at risk of poverty or underprotected, given his state in the previous period. If we know the individual's initial poverty position, a time path of poverty states can be constructed for each individual.

Policies for social inclusion such as activation, school dropout prevention and stimulating social participation will alter the transition matrix and thus also the time path of poverty states. To demonstrate the effect, at the beginning of our observation period (1 January 1993) we define target groups of respondents eligible for these measures. For school dropout prevention, we select all respondents who did not finish upper secondary education (including those who did not have lower secondary education). Simulation consists in assigning the corresponding level of education to these respondents. In the case of social capital building, people with a level lower than the mean MCA scores are taken as the target group. Simulation consists in enhancing their social capital by two standard deviations. The difference in the mean time path of poverty with and without simulation is considered to be the effect of the policy measure (De Blander & Nicaise, 2009).

In the case of activation, the target group comprises respondents who were out of work for at least three months after 1 January 1993. The simulation consists in "giving these people work" during 12 months. Note that for calculating the time path of poverty, the procedure followed here is slightly different from the other simulations in order to take into account the dynamic character of labour market participation. First the transition matrix of work is calculated and then it is integrated into the poverty path matrix. For a formal expression of this model we refer to annex 2 to this paper.

3.3. ADDING CHILDREN TO THE PICTURE

Children are defined as individuals aged less than 12 years. This rather low age limit avoids early exit from the sample due to leaving the parental home. To convert the adult poverty risks into their children's risks, we construct a re-weighted transition matrix for children by multiplying the adult transition matrix by the number of children per adult in each household. This is done with observed and simulated data to evaluate the potential effect of general policy intervention on the dynamics of child poverty. For a formal expression of this model we refer to annex 2 to this paper.

4. RESULTS

4.1. POVERTY OUTFLOW, EDUCATION, SOCIAL CAPITAL AND EMPLOYMENT

In Figure 3 (Annex 3) an average poverty path is constructed for respondents who are poor at the beginning of the observation period (1 January 1993). At this point, their poverty risk per definition equals 1. Hence, the dotted curve shows adults' mobility out of poverty. The corresponding (greyish) curve for children diverges from the adults' curve for three reasons: (a) adults with children have characteristics that distinguish them from other adults, resulting in different opportunities and transition probabilities, (b) households with (more) children face a higher poverty risk due to their family burden; and (c) the larger households get a larger weight (than single-child households) in the latter curve.

We observe that within 12 months, about 60% of initially poor individuals rise above the poverty line. Subsequent outflow is much slower, but in the end just a quarter of the initial subsample remains poor. The outflow of children is slightly lower than the adult poverty outflow. After twelve months, the curves diverge about 5 percent and this difference is sustained throughout the whole observation period.

Figure 4a and b (Annex 3) reflect specific outflow patterns in the cases of unemployment, low education and low social capital, respectively. A lack of human capital appears to be the most obvious disadvantage in escaping poverty. For individuals with less than lower secondary education the total outflow stagnates below 60%. Low social capital also implies a below-average outflow, while for individuals without upper secondary education it is only marginally smaller than for the rest of the population. For non-working individuals, exit is just delayed (cf. Fig. 3).

With regard to child poverty, we observe that its persistence is highest for children of parents with less than lower secondary education; their long-term outflow rate fluctuates around 40%. Low social capital results in an overall outflow just below 60%. The same applies to children with non-working parents, but in addition outflow is much slower. After 40 months the probability of escaping poverty barely exceeds 40%. The outflow pattern of children whose parents have a degree of lower secondary education is very close to the average pattern (cf. Figure 3).

Note that the effect of non-employment on adult poverty and child poverty is very different. While it is not a decisive factor hampering the average adult's mobility out of poverty, parents' employment status seems to be a determining element for children's mobility chances. Lack of (lower secondary) education is an important obstacle to adult mobility, but living with poorly educated parents seems to be an even greater barrier for children. For low social capital and lack of upper secondary education, the difference between adult poverty and child poverty outflow is much more similar.

4.2. THE POTENTIAL IMPACT OF POLICY MEASURES

What would happen to mobility if individuals were to achieve lower or upper secondary education, or an acceptable level of social capital, or were offered work during a period of one year? The curves at the bottom of each figure represent the difference between the actual outflow and the results of this simulation, indicating the potential impact of policy measures.

In the short run, granting people access to the labour market lifts almost everyone out of poverty, children even more than adults. However, after a year the activation effects begin to diminish. Some of the activated individuals slide back into poverty, while in the medium and long term the outflow out of poverty is substantial even without activation measures. After eight years of observation, the net impact of temporary subsidised work tends towards zero.

Enhancing people's educational level, on the other hand, does have a lasting impact on mobility out of poverty (Figure 4). If individuals without lower secondary education were to achieve this level, after eight years about 17% fewer would stay poor. The additional outflow induced by upper secondary education is just above 10%. The reduction in child poverty attributable to these measures is practically the same. In other words, preventing school dropout would lead to shorter poverty spells, for adults as well as children. Outflow is also notably enhanced through an improvement in individuals' social capital, leading to a similar reduction in child poverty – about 12%.

The irregularities in the child poverty curves are due to data imperfections. In addition, towards the end of the observation period, the number of children decreases since many have grown older than 12 while less newborns have entered the sample.

5. DISCUSSION

Using longitudinal modelling, in this paper we assessed differences in poverty dynamics between adults and children, focusing on poverty persistence and outflow patterns. Our study showed that mobility is large: over time a majority of the poor move out of poverty. This is slightly more the case for adults than for children, suggesting that the presence and the number of children in the household imply additional barriers to mobility. We should nevertheless bear in mind that a substantial minority of the poor *do* face long-term poverty (Muffels et al., 1998; Snel et al., 2000); in our study, a quarter of poor adults and even more children did not escape poverty over a period of eight years.

5.1. POVERTY DYNAMICS, EDUCATION, SOCIAL CAPITAL AND EMPLOYMENT

We situated these poverty dynamics within the recently developed theory of individualisation and destandardisation of the life course, turning poverty experiences into a more temporal and generalised phenomenon that is less dependent on social background. People are supposed to construct and be responsible for their path towards social inclusion. Nonetheless, not everyone has the same capacity to do so. We therefore observed considerable inequality in terms of social inclusion dynamics related to individual's resources such as education and social capital. We found that a lack of education, especially lower secondary education, and low social capital turn poverty into a more long-term phenomenon.

Parents are not only responsible for their own but also for their children's social inclusion. This complex task of balancing care and financial responsibility makes the pathway to social inclusion more difficult, especially when educational and social capital is limited. Low levels of human and social capital reduce the range of opportunities in the labour market *and* with respect to childcare, and they limit parents' planning and decision making skills. We found that especially when parents lack lower secondary education, child poverty becomes particularly persistent. Where social capital is low, there is no difference between general and child poverty outflow patterns.

The effect of non-employment on overall and child poverty also illustrates the burden implied by the presence of children. Although we found that unemployment only marginally prolongs poverty for the average adult, it has a considerable negative effect on children's poverty outflow rates. Whereas the average adult unemployment is often temporary or can be compensated through other income sources, for households with children it is a more persistent threat and financial deficiency seems to be harder to overcome.

5.2. THE IMPACT OF GENERAL ANTI-POVERTY MEASURES ON THE PERSISTENCE OF CHILD POVERTY

At first sight, it seems that general policy measures focusing on school dropout prevention and labour market participation would be most effective in tackling child poverty, since those are the most important risk factors for its persistence. Nevertheless, our research showed that activation is only partially effective in this context. Activation provides an immediate way out of poverty, shortening the period of poverty, but in the long run it has no effect on poverty reduction. This is due to the fact that a large proportion of those activated would have escaped poverty without additional intervention, while some slide back into poverty after the activation period. If labour market dysfunctionings cause people to lose their jobs in the first place, merely activating them by providing a temporary job will not necessarily diminish their difficulty in keeping it. To be effective in the long term, activation needs to be accompanied by capacity-building beyond the activation period. For parents to be able to keep a job, opportunities are needed to combine work with childcare: in addition to inclusive jobs, childcare facilities must be accessible and compatible with the working conditions of the persons activated. Note, however, that in the case of child poverty, the reduction of the poverty experience may also reduce the risk of poverty at later ages.

Since schooling concentrates explicitly on capacity-building, it was not surprising to find that improved educational outcomes have a considerable impact on sustainable poverty outflow. Our results also call for more investment in social organisations and social participation, since we found that social capital enhancement had a considerable impact on poverty outflow. Resources drawn from people's social networks are a useful helping hand out of poverty. The effects of the two latter policies (education as well as social capital) were not larger on general than on child poverty.

6. CONCLUSION

Poverty persistence, factors hampering outflow such as unemployment, low educational attainment and reduced social capital, and the effects of policy measures are different for adults and children. In an individualised society, where emphasis is placed on people's responsibility for their own social inclusion, the double responsibility of care and work becomes a particularly high burden on parents experiencing poverty. This leads to more persistent child poverty especially when parents lack educational resources. Education therefore appears to be the best policy tool to lift households sustainably out of poverty. Nevertheless, we must also acknowledge that none of the examined policies (activation, education or social participation) is able to bridge the gap completely between child and general poverty.

ANNEX 1

Figure 1. Multiple correspondence analysis of indicators of "bonding" and "bridging" social capital

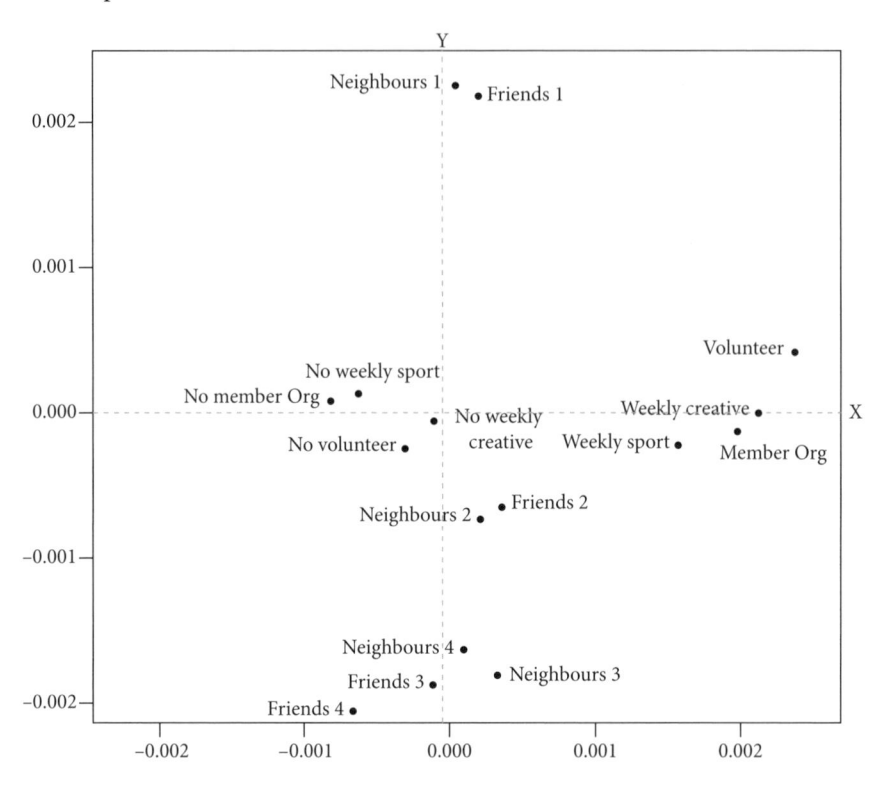

Table 1. Results of the dynamic estimation of poverty, based on key variables work, education and social capital

Trichotomised income	Non-poor			At risk of poverty			Underprotected		
	Coëff	St. error	P	Coëff	St. error	P	Coëff	St. error	P
Employment	0.312	0.027	0.000	0.164	0.040	0.000	0.566	0.082	0.000
Obs. = 306513; Log pseudolikelihood = –31388.36642									
Lower secondary	0.054	0.036	0.137	0.065	0.053	0.110	0.392	0.109	0.000
Upper secondary	0.133	0.042	0.001	0.143	0.061	0.010	0.476	0.131	0.000
Higher education	0.332	0.057	0.000	0.276	0.084	0.000	0.744	0.182	0.000
Obs. = 294832; Log pseudolikelihood = –31280.78331									
Bridging social capital	0.295	0.132	0.025	0.252	0.134	0.040	0.311	0.139	0.020
Obs. = 243981; Log pseudolikelihood = –25256.24582									
Bonding social capital	0.169	0.112	0.131	0.191	0.113	0.047	0.148	0.117	0.280
Obs. = 243981; Log pseudolikelihood = –25226.11954									

ANNEX 2

Based on these estimations, we calculate for each individual and for each month of observation the chance to be non-poor, at risk of poverty or underprotected. For the initial period t_1, this probability depends on the value on the key variables and on the control variables (including the correction terms). For all consecutive periods it depends in addition on the poverty state in the previous period.

$P(T)it_1 = \beta Si + \alpha Xi + \varepsilon$ *(initial estimation)*
$P(T)it = \beta Si + \beta'Tit\text{-}1 + \alpha Xi + \varepsilon$ *(dynamic estimation)*

where S stands for the key variable concerned, T for the poverty state, X for control variables, β, β' and α for the regression coefficients, i for the respondent, t for the time period and P(T) for the estimated poverty state.

With this last equation, a transition matrix (Mit) can be constructed indicating the probability for an individual to be in each of the poverty states (non-poor, at risk of poverty or underprotected), given the respondent's poverty state in the previous period.

$P(T)it \mid A_1 it\text{-}1$
$P(T)it \mid A_2 it\text{-}1 \quad = Mit$
$P(T)it \mid A_3 it\text{-}1$

where A_1 is the state "non-poor", A_2 is the state "at risk of poverty" and A_3 is the state "underprotected".

If, then, we know the individual's initial poverty position, a time path of poverty states can be constructed as follows:

$Tit = Tit\text{-}1 * Mit$

Poverty measures such as activation, preventing school dropping out or stimulating social participation will alter the transition matrix and thus also the time path of poverty states. To demonstrate the effect, at the beginning of our observation period (1 January 1993), we define target groups of respondents eligible for these measures. For school dropout prevention, we select all respondents who did not finish upper secondary education and those who did not have lower secondary education. Simulation consists in assigning the corresponding level of education to these respondents. In the case of stimulating social capital, people with a level lower than the mean MCA scores are taken as the target group. Simulation consists in elevating their social capital by 2 standard deviations. The difference in the mean time path of poverty with and without simulation can be considered to be the net effect of the policy measure (De Blander & Nicaise, 2009).

In the case of activation, the target group is built up of respondents without work for at least three months after 1 January 1993. Simulation consists in giving these people work during our first year of observation. Note that for calculating the time path, the procedure is slightly different in order to take into account the dynamic character of labour market participation. This implies that first the transition matrix of work is calculated and integrated in the transition matrix of poverty path.

On the basis of the transition matrix of work,

$P(W)it \mid W_0it\text{-}1$
$P(W)it \mid W_1it\text{-}1 \quad = M'it$

where W_0 is the state "not working" and W_{21} is the state "working".

And if we know the individual's initial work position, a time path of employment states can be constructed.

Eit= Eit-1 * M'it

The time path of poverty can then be constructed as follows:

Tit = Tit-1 * (Mit (w=1) * Eit;1 + Mit (w=0) * Eit;0)

Mit is calculated twice, once given that the individual i is working (Mit (w=1)) and once given that he is not (Mit (w=0)). The final transition matrix for poverty is the mean of both, weighted for the respective chances of being employed or unemployed.

ADDING CHILDREN TO THE PICTURE

Children are defined as individuals less than 12 years old at the beginning of our observation period. This age limit was chosen to avoid early exit from the sample due to leaving the parental home. To convert the adult poverty risk into their children's risk, we first construct a transition matrix for children, multiplying the adult transition matrix by the number of children per adult in the household:

Cit = Mit * (c/n)

with c/n the number of children per adult in the household.

The time path of child poverty can then be constructed as follows:

$T'_{it} = T'_{it-1} * C_{it}$

The mean time path for children is then:

$\check{T}'_t = \Sigma\, T'_{it} / (c/n)$

C_{it} can be calculated on the basis of M_{it} for observed data and for simulated data to evaluate the potential net effect of general policy intervention on the dynamics of child poverty.

ANNEX 3

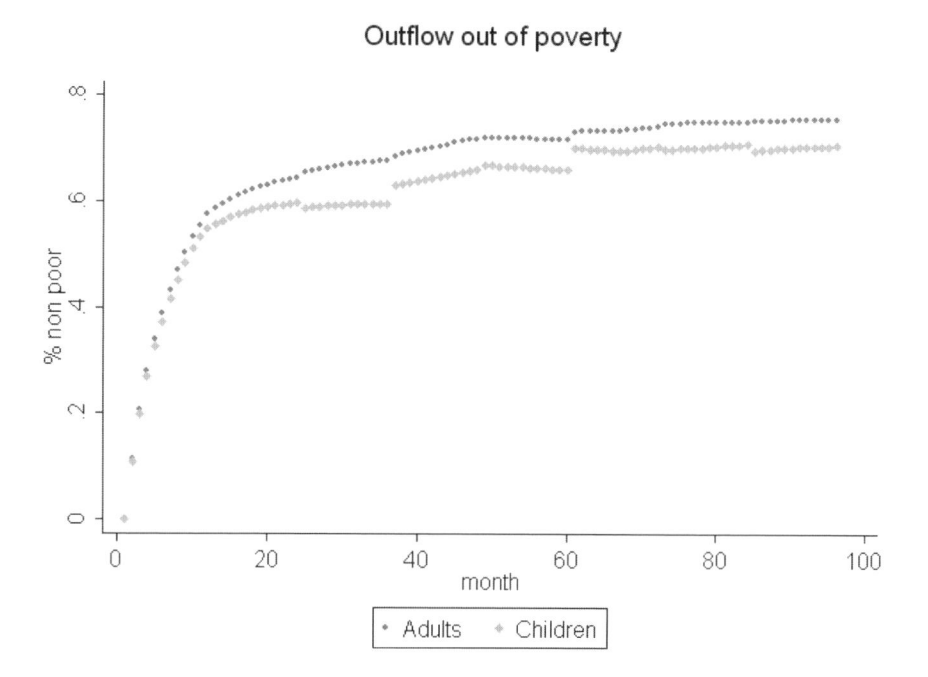

Figure 3. Total outflow out of poverty for adult respondents and for children over a period of 96 months between 1 January 1993 and 31 December 2001

Outflow out of poverty

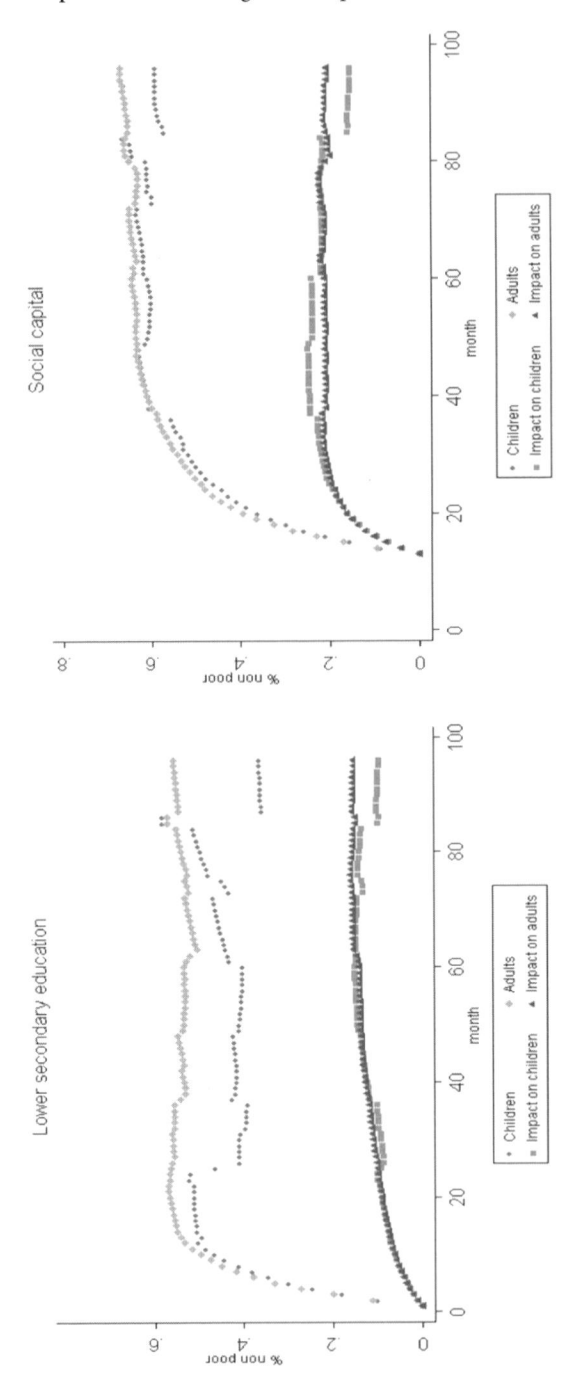

Figure 4a. Outflow out of poverty for adults and for children over a period of 96 months between 1 January and 31 December 2001 when lacking lower secondary education or social capital, and the influence of policy measures preventing school dropout and enhancing social capital formation

8. Escaping poverty with your children

Figure 4b. Outflow out of poverty for adults and for children over a period of 96 months between 1 January and 31 December 2001 when temporary unemployed or lacking upper secondary education, and the influence of labour force activation and policy measures preventing school dropout

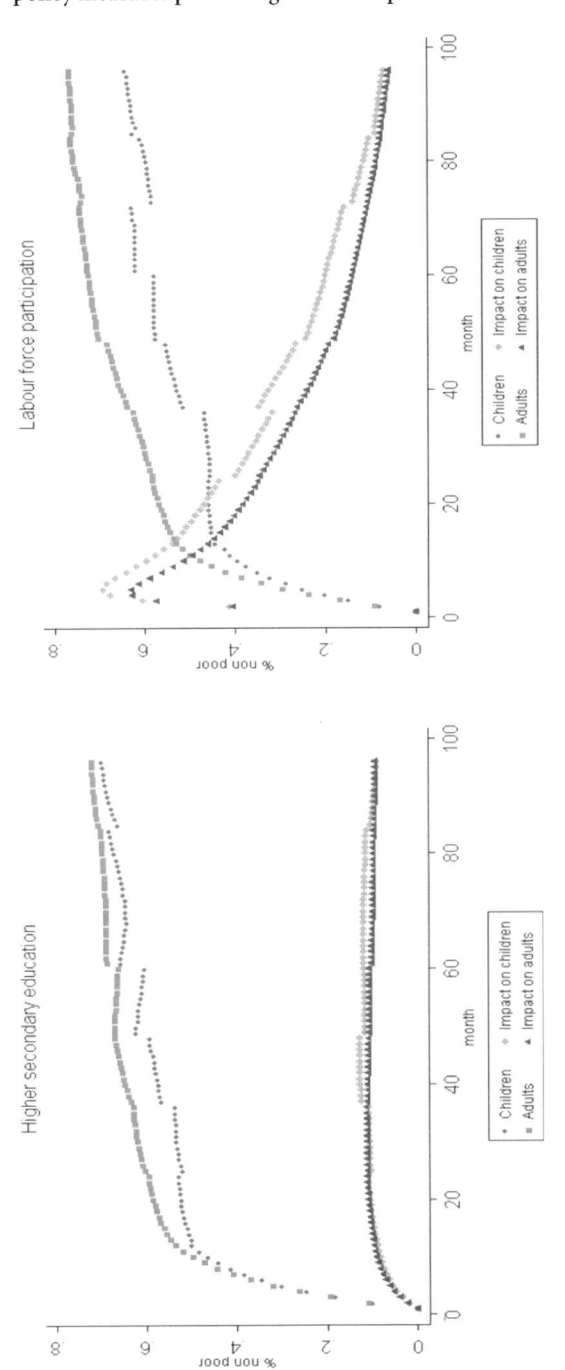

REFERENCES

AYALA L. & RODRIGUEZ M. (2006), The Latin model of welfare: Do 'insertion contracts' reduce long term dependence?, *Labour Economics*, 13, 799–822.

BANE M.T. & ELLWOOD D.T. (1986), Slipping Into and Out of Poverty: The Dynamics of Spells, *Journal of Human Resources*, 21, 1, 1–21.

BECK U. (1992), *Risk society. Towards a new modernity*, Sage, London.

BLUNDELL R., DEARDEN L., GOODMAN A. et al. (2005), Measuring the Returns of Education, in S. MACHIN & A. VIGNOLES (Eds.), *What's the Good of Education? The Economics of Education in the United Kingdom*, Princeton: Princeton University Press.

BOURDIEU P. (1993), *Sociology in Question*, Londen: Sage.

BRADBURY B., JENKINS S. & MICKLEWRIGHT J. (2000), *Child Poverty Dynamics in Seven Nations*, The Social Policy Research Centre, Australia.

COCKX B. & RIDDER G. (2001), Social employment of welfare recipients in Belgium: An evaluation, *Economic Journal*, 111, 470, 322–352.

COLEMAN J. (1988), Social capital in the creation of human capital, *American Journal of Sociology*, 94, 95–120.

DAHL E. & LORENTZEN T. (2005), What Works for Whom? An Analysis of Active Labour Market Programmes in Norway, *International Journal of Social Welfare*, 14, 86–98.

DAL G.A. (2007), Vers un Etat social 'actif'?, *Journal des tribunaux*, 127, 8, 133–139.

DE BLANDER R. & NICAISE I. (2005), De impact van maatschappelijke keuzen op de armoededynamiek, *Belgisch Tijdschrift voor Sociale Zekerheid*, 2005, 4, 675–709.

DE BLANDER R. & NICAISE I. (2009), Shooting at moving targets. Short versus long term effects of Anti-Poverty Policies, in: A. ZAIDI et al. (Eds.), *New frontiers in Microsimulation Modelling*, Burlington: Ashgate, 471–198.

DE BOYSER K. & LEVECQUE K. (2007), Armoede en sociale gezondheid: een verhaal van povere netwerken?' in: J. VRANKEN et al. (Eds.), *Armoede en sociale uitsluiting. Jaarboek 2007*, Antwerpen: Acco/Oases, 167–177.

DEWILDE C. & LEVECQUE K. (2003), Armoededynamieken herbekeken – Het verhaal achter het verhaal, in: J. VRANKEN et al. (Eds.), *Armoede en sociale uitsluiting. Jaarboek 2007*, Antwerpen, Acco/Oases, 229–243.

DIELEMAN A. (2000), Individualisering en ambivalentie in het bestaan van jongeren, *Pedagogiek*, 20, 2.

DUBIN J.A. & MCFADDEN D.L. (1984), An econometric Analysis of Residential Electric Appliance Holdings and Consumption, *Econometrica*, 52, 2, 345–362.

FPS ECONOMY, SMES, MIDDLE CLASSES & ENERGY (2010), *Statistics on Income and Living Conditions: Poverty*, http://statbel.fgov.be/nl/statistieken/cijfers/arbeid_leven/inkomens/armoede/index.jsp.

GROENEZ S., HEYLEN V. & NICAISE I. (2010), *De opbrengstvoet van investeringen in het hoger onderwijs: een verkennend onderzoek*, Leuven: HIVA-K.U.Leuven.

HECKMAN J.J. (1979), Sample Selection Bias as a Specification Error, *Econometrica*, 47, 1, 153–161.

HERMANS K. (2005), De actieve welvaartstaat in werking: OCMW's en het activeren van leefloongerechtigden, in: J. VRANKEN et al. (Eds.), *Armoede en sociale uitsluiting. Jaarboek 2002,* Antwerpen: Acco/Oases, 179-202.

HOOGHZ M. (1999), Inleiding: verenigingen, democratie en sociaal kapitaal, *Tijdschrift voor Sociologie,* 20, 3-4, 233-246.

LEISERING L. & LIEBFRIED S. (1999), *Time and Poverty in Western Welfare States, United Germany in Perspective,* Cambridge: Cambridge University Press.

LORENTZEN T. & DAHL E. (2005), Active labour market programmes in Norway: are they helpful for social assistance recipients?, *Journal of European Social Policy,* 15, 27-45.

MACHIN S. & MC NALLY S. (2006), *Gender and student achievement in English schools,* London: Centre for the Economics of Education, London School of Economics and Political Science.

MCINTOSH S. (2004), The impact of Vocational Qualifications on the Labour Market Outcomes of Low-Achieving School-Leavers, *CEP Discussion Paper dp0621,* Centre for Economic Performance.

MUFFELS R., SNEL E., FOUARGE D. et al. (1998), Armoedecarrières. Dynamiek en Determinanten van Armoede, in: G. ENGBERSEN, C. VROOMAN & E. SNEL (Eds.), *Effecten van Armoede. Derde Jaarrapport Armoede en Sociale Uitsluiting,* Amsterdam: University Press, 45-65.

MYLES J. (1993), Is there a post-fordist life course?, in: W.R. HEINZ (Ed.), *Institutions and gatekeeping in life course,* Weinheim: Deutchen studien verslag (Deutscher Studien Verlag), 171-185.

NICAISE I., GROENEZ S., ADELMAN L., et al. (2004), *Gaps, traps and springboards in European minimum income systems. A comparative study of 13 EU countries,* Leuven: HIVA-K.U.Leuven.

NICAISE I. (1998), Armoede en menselijk kapitaal, *De Gids op Maatschappelijk Gebied,* 89, 3, 231-244.

OECD (2005), *Education at a Glance,* Paris: OECD.

PEETERS B. & VRANKEN J. (2000), *Effectiviteit van hulp en zorg ten aanzien van de sociale en culturele integratie,* Antwerp: UFSIA-OASeS.

PORTES A. (1998), Social capital: its origins and applications in modern sociology, *Annual Review of Sociology,* 24, 1, 1-24.

PUTMAN R.D. (1995), Bowling alone: America's declining social capital, *Journal of Democracy,* 6, 1, 65-78.

SNEL E., ENGVERSEN G. & VROOMAN C. (2000), Modernized Poverty. Individualization, concentration and embeddednes, in: J. BERGHMAN, A. NAGELKERKE et al. (Eds.), *Social Security in Transition,* Den Haag: Kluwer Law International, 63-76.

STEENSSENS K., DEMEYER B., VAN REGENMORTEL T. (2009), *Conceptnota empowerment en activering in armoedesituaties,* Leuven: HIVA-K.U.Leuven.

STEENSSENS K., AGUILAR L.M., DEMEYER B. et al. (2008), *Kinderen in armoede. Status quaestionis van het wetenschappelijk onderzoek voor België,* Leuven: IGOA-GiReP, HIVA-K.U.Leuven.

STEENSSENS, K., VANDENABEELE, J., PULTAU, W. et al. (1996), *De netwerken van de armen*, Brussel: Federale Diensten voor Wetenschappelijke, Technische en Culturele Aangelegenheden.

STEUNPUNT VOOR BESTRIJDING VAN DE ARMOEDE, BESTAANSONZEKERHEID EN SOCIALE UITSLUITING (2008), *Prioriteiten voor het Belgisch Voorzitterschap van de Europese Unie*, www.armoedebestrijding.be/publications/nota_Belgisch_voorzitterschap_2010.pdf.

THYS R. (2001), *Van kansarme jongeren naar arme netwerken. Relationele micromechanismen van sociale uitsluiting*, Doctoral Thesis, University of Antwerp.

THYS R., DE RAEDEMAECKER W. & VRANKEN J. (2004), *Bruggen over woelig water. Is het mogelijk om uit de generatie-armoede te geraken?*, Leuven/Voorburg: Acco.

VENTURINI G.L. (2008), Poor Children in Europe: An Analytical Approach to the Study of Child Poverty in the European Union, Between 1994 and 2000, *Child Indicators Research*, 1, 4, 323-349.

WORLD BANK (1998), The Initiative of Defining, Monitoring and Measuring Social Capital: Overview and Program Description, *Social Capital Initiative Working Paper 1*, Washington D.C.: World Bank.

9. EARLY CHILDHOOD POVERTY IN THE EU: MAKING A CASE FOR ACTION

Katrien De Boyser

1. CHILDHOOD POVERTY AS A SOCIAL RESEARCH AND POLICY CONCERN

During the past decade, the fight against childhood poverty has become a growing issue in the social policies of the EU and many of its member states. In their most recent National Reports on Strategies for Social Protection and Social Inclusion, most member states put forward integrated strategies to prevent and alleviate childhood poverty and social exclusion. Taking on this major social problem through EU policy is also a central theme during 2010, the EU Year for Combating Poverty and Social Exclusion.

One mobilising fact is that in the EU children are at greater risk of living in poverty than adults. Childhood poverty affects the short-term well-being of children and their individual life courses. On a societal level, a high level of childhood poverty is also likely to result in substantial macro-effects, especially in the long run. These effects can essentially be looked upon as losses in terms of both economic and human capital, and can affect social cohesion. In turn, substantial societal benefit can be gained from investments aimed at preventing or remedying the effects of poverty in the earliest childhood years (Cleveland & Krashinsky, 2003; Duncan et al., 2008).

This macro perspective is also relevant when looking at childhood poverty from a sociological perspective, as there is a strong link to broader theories and research on social stratification and social mobility. As many researchers in this field take great interest in the inheritance of social positions over generations, the intergenerational transmission of poverty has been one of their central research issues in the past few decades. This is also where the intra-generational life-course perspective – the movements up and down the social ladder within generations – comes in. In poverty research, this results in questions concerning the extent and impact of periods of childhood poverty: how many children are poor, how do they live, what are the immediate effects of different types of deprivation, to what extent and how does experience of poverty and deprivation during childhood affect future chances of social mobility, how straightforward are the links between both, and which are the intermediary factors hindering or enhancing future mobility chances? Although many questions remain unanswered in this respect, there are interesting research findings to explore.

More and more evidence is, however, becoming available on the importance of the earliest life stage in establishing the crucial foundations for future achievement and social mobility. In this chapter, therefore, we want to focus on a selection of recent literature on the short- and long-term impact of childhood poverty in the earliest life stage. Secondly, as this volume is being published in the European Year for Combating Poverty and Social Exclusion, we give an overview of early childhood deprivation in the EU-27 and focus especially on the circumstances in which the youngest (zero to six year old) children are growing up – an area in which we are also faced with some measurement issues. In conclusion, we take a look at policy recommendations from international research, which may be relevant to EU policies on early childhood poverty and its outcomes.

2. WHY FOCUS ON POVERTY IN THE EARLIEST LIFE STAGE?

Before looking at a snapshot of early childhood poverty and deprivation levels in the EU-27, we want to show why it is important not to ignore this critical life phase in the EU policy debate on childhood poverty. Prevention of early deprivation is not often a subject of EU or national policies.

In the following paragraphs we will focus on this lack of attention to the earliest life stage in both research and policy and – since this may be critical – on the outcomes of early childhood poverty in terms of short- and long-term health, developmental and socio-economic effects. We adopt a multidisciplinary approach in this brief literature review, underpinned and explained by mutually reinforcing evidence from different perspectives.

2.1. IS EARLY DEPRIVATION TOO OFTEN A MISSING ELEMENT IN THE SOCIAL MOBILITY DEBATE?

Childhood is generally considered to be the most important life stage because it lays the foundations for future personal social and economic achievement. In this achievement process, educational attainment plays a central role as it is an important public instrument of upward social mobility. In spite of the post-war belief in the democratisation of European educational systems, and in spite of the general development of welfare states, upward social mobility has remained more difficult for those growing up at the bottom of the socioeconomic ladder. Going through the education system has not prevented the inheritance of social positions from persisting. Currently, this can mean that the most disadvantaged are potentially worse off, as human capital, often seen in terms of educational attainment, is considered a crucial production factor in EU knowledge economy (see both the original and revised Lisbon Strategy).

It was only rather recently that social researchers started pointing to the earliest life stage as a potentially crucial mediating factor in this social 'attainment' process. Some social scientists realised that we may have been looking at just a part of the puzzle when expecting democratic education systems to counter problematic social inheritance processes for children from low SES (socio-economic status) households. Or to quote Esping-Andersen (2004: 133): "The standard causal model that underpins most research, namely the 'origins → education → destiny' sequence is, if not mis-specified, at least seriously underspecified".

A recent longitudinal study of a 1953 birth cohort (Bäckmann & Nillson, 2010) clearly shows that resource deficiency during childhood has long-term effects on social exclusion risks, and identifies long-lasting periods of poverty in childhood as most detrimental for future achievement. This Swedish research confirms earlier findings in Northern America which established that people who lived their childhood in extreme poverty or faced long-lasting periods of poverty had a significantly higher risk of also living their adult lives in poverty. The effects of deprivation on future social mobility have also been found to be significantly stronger when they occurred in the earliest life stage than later on in childhood, for example in adolescence (Brooks-Gunn & Duncan, 1997). Looking for explanations as to why this stage is so crucial leads us to explore some findings from other research disciplines.

2.2. POVERTY, EARLY LIFE HEALTH AND HEALTH THROUGHOUT THE LIFE-COURSE

Many of the findings on the life-course effects of early childhood poverty stem from health and epidemiology studies. There is a vast body of evidence to show that the poor socio-economic characteristics of the household(s) and environment(s) in which children grow up have both short- and long-term effects on their well-being in terms of health and on other life domains (see amongst others Blackwell et al., 2001; Chase et al., 1997; Kestila et al., 2005).

In the short run, the early child health consequences of poverty are multiple. From pregnancy onwards, mothers with low socio-economic status are, more often than other mothers, faced with stressful life events (such as single-parent or teenage pregnancy, unemployment, crowded or polluted physical environments) and have fewer resources and coping mechanisms to deal with these risks (Larson, 2007). It is generally understood that in the EU there are still significantly higher perinatal health (morbidity) and mortality risks for children born into more disadvantaged socioeconomic households than for other children. The relationship between mothers' low socio-economic status and infant mortality, low birth weight and preterm birth is especially clear. Administrative data in Flanders show, for example, that – despite the generally very low figures – perinatal mortality risks are still strongly associated with socio-economic status in terms of parental educational attainment and job status. Next to socio-economic background, the pregnancy process and birth weight play an important role (Vlaams Agentschap Zorg en Gezondheid, 2007). On top of the risk of already having a worse health status at birth, children in poverty are often exposed to additional health risks such as unhealthy housing conditions and poor nutrition, which can worsen their health status. Children growing up in a situation of severe income deprivation are more prone to chronic illnesses (such as asthma), retarded growth and hospitalisation (Se'guin et al., 2005; Kestilla et al., 2005).

Moreover, the effects are also visible in the long run. In fact, a major focus of life-course epidemiology has been to understand how early life health experiences shape adult health (Braveman & Colleen, 2009). There is much epidemiological evidence that early childhood poverty affects long-term adult health both in terms of physical and mental health and in terms of health behaviour. The effects are especially well documented for cardiovascular conditions, cancer, lung diseases and arthritis/rheumatism (see amongst others Blackwell et al., 2001). Haas (2007) showed that these early childhood factors not only influence the overall level of adult health outcomes but also the trajectories of adult health over time.

2.3. ON EARLY CHILDHOOD POVERTY AND DEVELOPMENT: IS POVERTY A BRAIN DRAIN?

Early childhood experiences and circumstances are important in terms of future health outcomes and well-being, and are fundamental for future learning processes and behaviour. As developmental research has shown, a highly complex process of cognitive, social and emotional development takes place in the earliest life stage. Aptitude for later developmental and learning processes, which are crucial in terms of educational attainment, is shaped during this early life stage (Sapolsky, 2004; Shonkoff & Phillips, 2000). Children growing up in poverty are disproportionately often exposed to factors that may hinder these developmental processes (Shore, 1997). More than other children, they are likely to experience psychological stress induced by stressful events of high frequency and intensity, often on a daily basis. Moreover, financial unpredictability, family instability, limited social support, exposure to damaging substances in the environment (such as fungi and lead), and limited cognitive stimulation all prove to have a potentially negative impact on cognitive, social and emotional development. It is especially the cumulative risk exposure that seems to form the greatest threat to the child's development[1] (Brooks-Gunn & Duncan, 1997). There is evidence that early deprivation – similar to adverse childhood living conditions increasing the susceptibility to general morbidity – also affects cognition and cognitive decline later in life (Fors et al., 2009; Luo & Waite, 2005).

The rapid behavioural and cognitive development of children in the first years of life – at the age from zero to three they learn to play, talk, walk, reflect, 'listen'… – is a reflection of the critical brain development processes which take place in early life. Socio-economic disadvantages during pregnancy and in early childhood can hamper the early, crucial development of brain structures (Evans, 2004). The development of the central nervous system can be threatened by different environmental factors such as poor nutrition, toxins and chronic stress, which are all more likely to be present in situations of poverty (see amongst others Shonkoff, 2000).

3. EARLY CHILDHOOD POVERTY AND DEPRIVATION IN THE EUROPEAN UNION

As mentioned above, research findings show that it could be crucial for EU social policy making, and for other policy domains, to gain a better and more focused understanding of the circumstances in which very young children grow up. In this section we will therefore map deprivation and poverty in this earliest life stage in the EU-27 and will look at some new childhood poverty measurement issues currently confronting the EU.

A first glance at the situation in the enlarged European Union (Figure 1) reveals that the overall at-risk-of-poverty rate was 16.2% in 2006 (EU SILC,[2] 2007). This

---

[1] Research in the US showed that there was a substantial difference in cognitive stimulation according to parental SES in terms of the level of interaction between children and parents, the richness of vocabulary, reading, and the level of PC and internet access (Evans, 2004).

[2] The European Union Survey on Income and Living Conditions is the main source for the compilation of comparable indicators on social cohesion used for policy monitoring at EU level in the framework of the Open Method of Coordination. It provides on an annual basis timely and comparable multidimensional micro-data on income, poverty, social exclusion and living conditions. Every year, both cross-sectional data (pertaining to a given time or a

means that about one in six people in the EU were living in a household whose equivalised income was below 60% of the national median equivalised income.³ The risk of living in a poor household in the EU-27 is in general higher for children (0–17 years old) than for adults: about one in five (19.1%) children are growing up in relative income poverty; for the youngest age group (0 to 6 years old) this is 17.7%. When looking at early childhood poverty purely from a relative income deprivation perspective, countries with the highest early childhood poverty levels are the United Kingdom, Italy, Poland, Spain and Luxemburg (which may be an odd case, as we will see later). In general, the Scandinavian countries have the lowest early childhood poverty levels, together with a number of new member states such as the Czech Republic and Estonia.

Figure 1. At-risk-of-poverty levels (%) in the EU-27, early childhood (0 to 6 years old) and all ages, EU SILC, 2007 (income 2006)

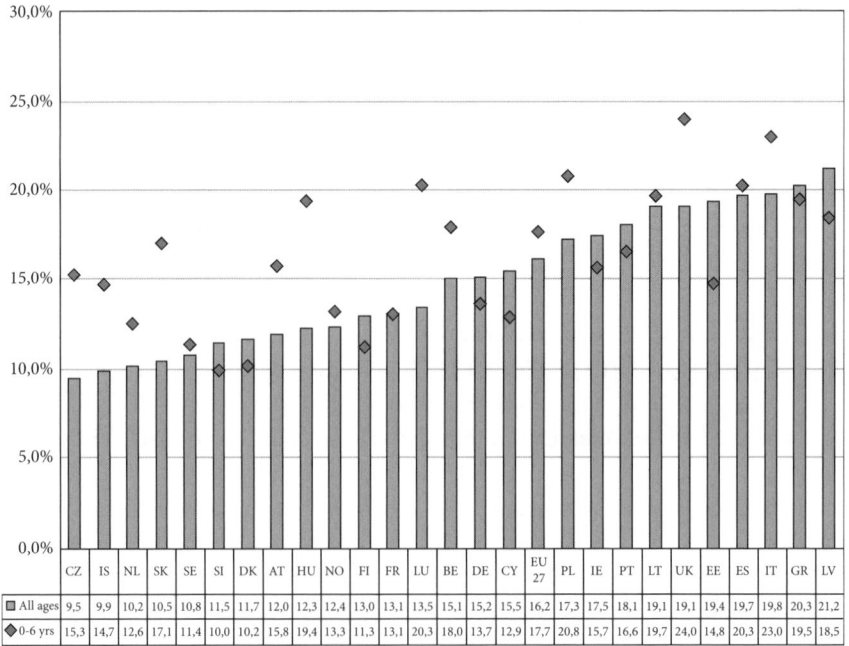

	CZ	IS	NL	SK	SE	SI	DK	AT	HU	NO	FI	FR	LU	BE	DE	CY	EU 27	PL	IE	PT	LT	UK	EE	ES	IT	GR	LV
All ages	9,5	9,9	10,2	10,5	10,8	11,5	11,7	12,0	12,3	12,4	13,0	13,1	13,5	15,1	15,2	15,5	16,2	17,3	17,5	18,1	19,1	19,1	19,4	19,7	19,8	20,3	21,2
0-6 yrs	15,3	14,7	12,6	17,1	11,4	10,0	10,2	15,8	19,4	13,3	11,3	13,1	20,3	18,0	13,7	12,9	17,7	20,8	15,7	16,6	19,7	24,0	14,8	20,3	23,0	19,5	18,5

Source: EU SILC 2007 (ADSEI), author's own calculations.

> certain time period) and longitudinal data (pertaining to individual-level changes over time, observed periodically over, typically, a four-year period) are collected.
>
> 3 Considered is the total household income (including the earnings of all household members, social transfers received by individual household members or the household as a whole, capital income…). Household income is equivalised on the basis of the OECD modified equivalence scale in order to take account of the differing needs of households of different size and composition. The OECD modified equivalence scale assigns a value of 1 to the first adult in the household, 0.5 to each other adult, and 0.3 to each child below the age of 14.

9. Early childhood poverty in the EU: making a case for action

Even though the above at-risk-of-poverty measure has been the most prominent social inclusion indicator in the EU for the last decade, scientists and policy-makers are currently rightly questioning its legitimacy as the prime indicator given the new enlarged EU context (Whelan & Maitre, 2009; SPC, 2008). Even though a number of new member countries have low general living standards, their exposure to income poverty seems to be similar to older and 'richer' member states, implying that this indicator only reflects part of the total picture. In order the better to express differences in living conditions – and not just a narrow income distribution – we will look at other resources and outcomes which can better assess different aspects of poverty and deprivation.

3.1. ECONOMIC DEPRIVATION AND LACK OF DURABLES

In order to capture the economic deprivation of children under six years of age in different EU member states, we use items from the EU SILC capturing different economic characteristics of the households in which they are living. From the EU SILC 2007 database, we selected four items representing economic deprivation: (1) having had arrears in the last twelve months on mortgage, rent payments, utility bills or loans; (2) not being able to afford a one-week annual holiday away from home; (3) not being able to face unexpected financial expenses; and (4) having difficulties making ends meet. In order to make a cross-national comparison, we built a summary measure that defines a person as being economically deprived when the household they live in is deprived in respect of at least two items. The EU SILC also put questions to measure a lack of durable goods because of financial restraint in the household: listed are lack of a TV, washing machine, computer, car and telephone in the home because the household cannot afford them.

In figure 2 we can see the at-risk-of-poverty rates for the households in which children under six are living, and added to this the two constructed measures described above: economic deprivation and lack of at least one of the listed durable goods because of inadequate resources. This picture already looks quite different: whereas the relative income poverty measure is rather similar for many EU countries, the measures of economic deprivation and lack of durable goods show a greater economic divide between new member states, together with 'old' southern member states such as Portugal, and the older, often more affluent member states. Whereas in the Slovak Republic, for example, the relative poverty rate for the youngest group (17%) is under the EU average (17.7%), one third (32%) of households lack at least one of the listed durable goods and three fifths (59.3%) of the youngest children live in households experiencing at least two forms of economic deprivation. Other countries with high levels of small children living in economically deprived households are Hungary, Latvia,

Poland, Lithuania, Cyprus, Portugal, Ireland, the Czech Republic, Italy and Greece. Countries with the lowest prevalence of economic deprivation are Denmark, Norway, the Netherlands and Sweden. Luxemburg, contrary to the high at-risk-of-poverty levels for children, is doing well on this indicator, a situation which is probably related to the narrower income distribution in Luxemburg.

Figure 2. At-risk-of-poverty levels (%), levels of deprivation of a minimum one of the listed durable goods and a minimum two types economic deprivation in early childhood (0 to 6 years) in the EU-27, EU SILC 2007 (income 2006)

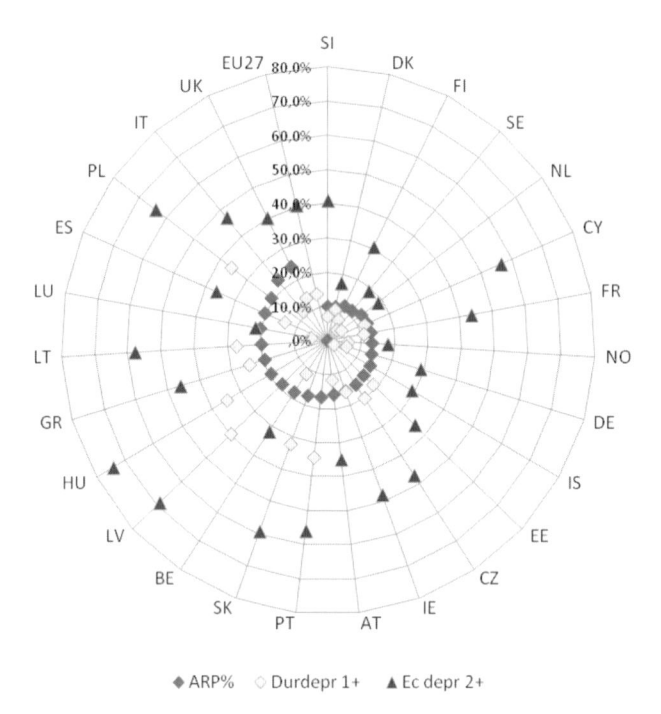

Source: EU SILC 2007 (ADSEI), author's own calculations.

3.2. NUTRITIONAL DEPRIVATION

Another interesting item in the EU SILC is whether the household can afford a meal with meat, chicken or fish (or a vegetarian equivalent) every second day. As this can be considered a grave and rather absolute effect of economic deprivation, we treat it separately from the items indicating forms of financial deprivation as such. In figure 3, we can see that the general rates of nutritional deprivation also reflect the economic divide between richer old and poorer new (and old) member states better than the general income poverty rate. Let us, for example, compare Belgium and Hungary, where children aged zero to six have a similar at-risk–of-poverty rate (18% and 19.4%): in Belgium this is not reflected in the proportion of children living in families experiencing food deprivation (2.7%), whereas for Hungary one in four small children are in this situation.

Figure 3. At-risk-of-poverty rate (%) and percentage of children (0 to 6 years) living in households not able to afford a meal with meat, chicken or fish (or a vegetarian equivalent) every second day in the EU-27, EU SILC 2007

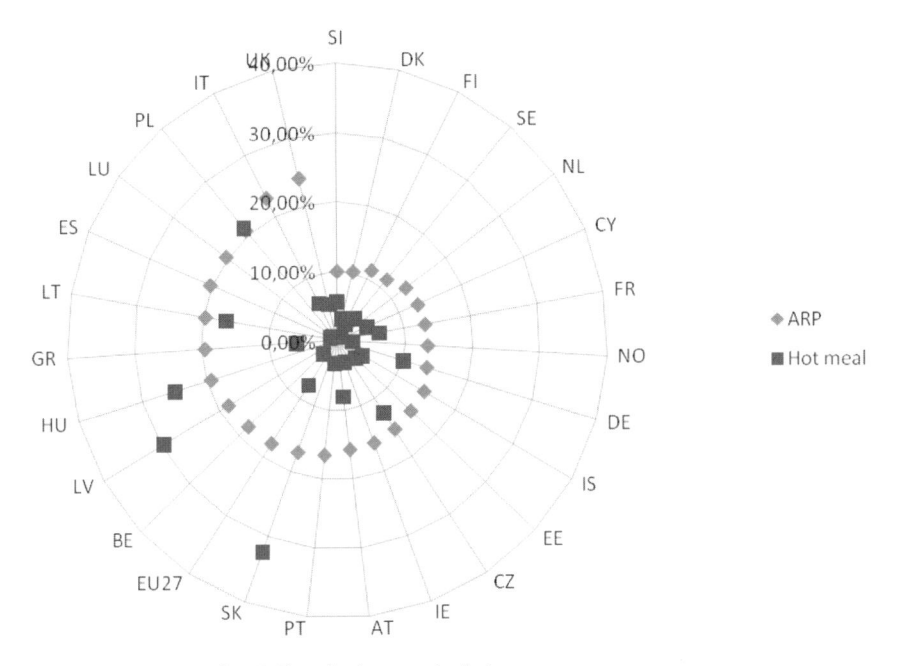

Source: EU SILC 2007 (ADSEI), author's own calculations.

3.3. HOUSING DEPRIVATION

Another important factor in the general well-being and health of children is their housing environment. In order to grasp housing deprivation, four items from the EU SILC were selected in order to construct a measure: (1) the house has a leaking roof, damp walls/floors/foundations, or rot in the window frames; (2) there is no bath or shower in the dwelling; (3) the household has no indoor flushing toilet for its sole use; and (4) the accommodation is too dark. Again, following a recent publication of the Social Protection Committee (2008), we established a cut-off at the occurrence of at least one of these types of housing deprivation. In general, more children in the EU-27 are found in one or more types of housing deprivation than there are children in households in relative income poverty. Figure 4 shows the disparities between housing deprivation and the at-risk-of-poverty rate. Again, the greatest differences between relative and 'more absolute' forms of deprivation are found in new EU member states, indicating as before that the pure relative income measure no longer stands its ground after EU enlargement.

Figure 4. At-risk-of-poverty rate (%) and percentage of children (0 to 6 years) in EU-27 households experiencing at least one of the following forms of housing deprivation: leaking roof, damp walls/floors/foundation, rot in window frames, no bath or shower, no indoor flushing toilet, dwelling too dark, EU SILC 2007

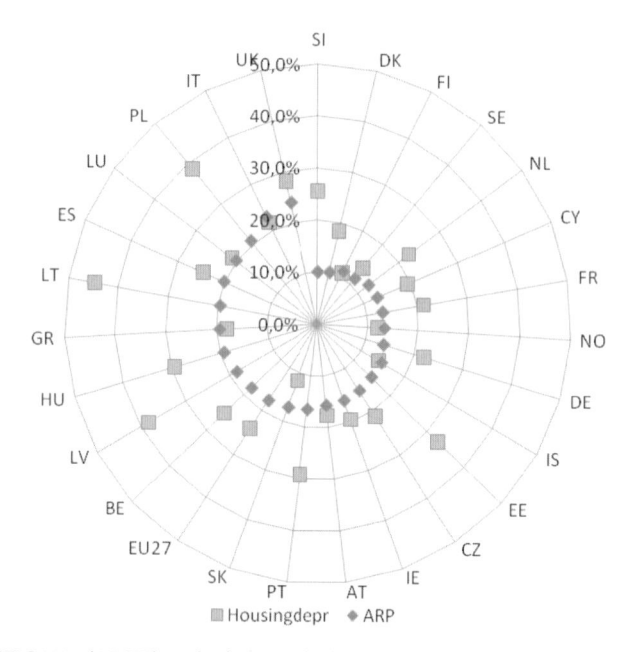

Source: EU SILC 2007 (ADSEI), author's own calculations.

3.4. MULTIPLE DEPRIVATION

Table 1 shows us different degrees of deprivation in early childhood in the EU-27. Next to the at-risk-of-poverty measure, we looked at the simultaneous presence of the constructed measures described above. This helps us to fine-tune our view of children in the EU at different degrees of deprivation. In the EU-27 in general, nearly one in three (29%) of children between 0 and 6 years old live in households experiencing economic deprivation, a lack of durable goods, food deprivation or housing deprivation; one in four (23.7%) experience at least two of these deprivation types; 8.5% at least three; and 1.6% experiencing deprivation in all defined dimensions. Latvia, Poland, Hungary, Lithuania, and the Slovak and Czech Republics score above average on the overall deprivation measure, whereas a country such as Luxemburg scoring high in the at-risk-of-poverty rate shows relatively low rates on the multiple deprivation measures.

Table 1. At-risk-of-poverty rate (%) and percentage of children (0 to 6 years) in households in the EU-27 experiencing at least one, two, three or four dimensions of deprivation, EU SILC 2007

	AT	BE	CY	CZ	DE	DK	EE	ES	FI	FR	GR	HU	IE
ARP	15.8	18.0	12.9	15.3	13.7	10.2	14.8	20.3	11.3	13.1	19.5	19.4	15.7
1 type depr	28.6	24.6	40.5	28.8	28.3	18.1	30.8	30.3	27.8	30.8	30.0	31.0	30.7
2 types depr	18.0	19.9	23.2	28.2	16.4	12.5	24.5	20.2	10.5	22.4	27.3	50.0	23.0
3 types depr	6.4	5.7	5.4	12.2	5.1	4.3	9.6	5.4	1.6	6.0	8.6	24.1	8.5
4 types depr	1.3	1.0	0.6	3.2	0.5	0.9	1.3	0.2	0.2	0.6	1.0	7.7	1.4

	IS	IT	LT	LU	LV	NL	NO	PL	PT	SE	SI	SK	UK	EU27
ARP	14.7	23.0	19.7	20.3	18.5	12.6	13.3	20.8	16.6	11.4	10.0	17.1	24.0	17.7
1 type depr	27.1	32.6	21.7	24.8	23.7	25.6	21.5	26.2	28.5	20.1	30.4	28.6	30.8	29.1
2+ types depr	9.7	23.0	47.4	10.8	54.5	9.6	7.7	49.2	39.3	8.7	21.1	41.2	23.8	23.7
3+ types depr	2.0	6.3	23.4	2.2	31.7	2.0	1.4	26.7	14.7	2.2	5.5	20.6	8.1	8.4
4 types depr	0.7	1.0	6.4	0.1	12.1	0.1	0.3	8.4	1.2	0.6	1.5	3.4	0.8	1.6

Source: EU SILC 2007 (ADSEI), author's own calculations.

4. THE EUROPEAN UNION AND EARLY CHILDHOOD POLICIES

The evidence of early deprivation having long-term effects on health, educational and socio-economic outcomes is large and growing, and seems to be just "bad news". This dark cloud nevertheless seems to have a silver lining, as early childhood policies can be efficient in reducing these effects.

A report from the European Social Protection Committee (2008) showed that the countries most successful in fighting child poverty tackle the problems in multiple ways and combine universal support for all children with targeted policy measures for the most vulnerable children. General and specific tax measures and parental income supplements can sustain household financial resources adequate for providing children with healthy houses and environments in which to grow up. Affordable, high-quality after-school and preschool child care services are found to be crucial by a number of EU member states in enabling parents to participate fully in the labour market and for achieving better early development for children living in deprived situations. In order to allow parental labour market participation, working time and leave measures must help to balance work and family life. Another element relevant for the earliest life stage that we found in the report is support for all parents in their parental role so that they are able to provide the best and safest environment for their children.

Working on socio-economic health disparities from the prebirth period onwards is another important policy recommendation found in much of the literature, although it remains often somewhat less apparent in national policies. The lasting health impact of poor health status in the perinatal period makes it particularly important to reduce the social and economic health inequalities associated with this period. For example, when looking at prenatal and perinatal health risks, policies can be developed to prevent prenatal exposure to poor maternal nutrition, infection and environmental neurotoxins (e.g. alcohol, lead); to provide systematic paediatric care for infants in order to screen for and treat many important causes of early developmental retardation; and to provide universal access to health care for pregnant women and children.

Researchers who have examined the long-term benefits of societal investment in the early childhood years have generally concluded that the benefits far outweigh the costs (Heckman, 2006; Duncan et al., 2008; Cleveland & Krashinsky, 2003; Rolnick & Grunewald, 2003). The eventual cost of waiting until educational and other arrears become "hard facts" is much higher than supporting poor families and children early on. And this early stage can be taken quite literally, or as Walfogel (2004: 1) puts it: "A good deal of inequality is already apparent by the time children start school, and children's development may be less amenable to change once they enter school".

Translating new research findings into ready-made policies is of course not always as straightforward. However, much can be learnt from different industrialised countries such as the Netherlands and New Zealand, where the early childhood poverty policy process has been running for some time. Many researchers also put forward tentative general guidelines for policy-makers who have to devise and implement poverty and other policies concerning children. Looking beyond the national borders, through researching and peer reviewing the principles of good practices there, can inspire member states to move forward on this important policy theme.

5. TO CONCLUDE: A STRONG CASE FOR ACTION

Childhood poverty has drawn the attention of different scientific fields, such as social and political science, psychology, educational science, economics, medicine and neuroscience, and human and children's rights. Research findings from these different angles have accumulated into a large knowledge base concerning the short- and long-term risks of growing up in poverty, both for the individual and for society, and in terms of future losses of both human and economic capital.

Although the earliest life stage often has not been adequately acknowledged by social researchers, it proves to be a quite important one when assessing social mobility processes in looking at the effects of parental SES on children's educational and socioeconomic attainment and at intragenerational social mobility. The short-term and long-term consequences of early childhood poverty are to be taken seriously: bad housing, poor health and other risk factors are found to have a potentially negative impact on health, brain development, educational attainment and general well-being. It is a positive token that tackling child poverty and breaking the transmission of poverty and exclusion features high on the European Union's political agenda.

Looking at the extent of the problem in the EU, we faced some methodological issues. After the enlargement of the European Union to 27 member states, the most cited EU poverty measure seems to be less adequate in comparatively measuring deprivation than before. In this chapter we used – next to the at-risk-of-poverty rate – multiple deprivation measures reflecting, among others, poor housing conditions and economic strain in households with children under the age of six.

We conclude this chapter by looking at general policy recommendations on early childhood poverty. Existing knowledge in neuroscience, psychology, economics, sociology and other fields can give important guidance to policy. Generally they suggest that the earliest years of life may be an important and promising time to make an impact on the lives of poor children. In answering the question how to intervene, it is important also to take the children's rights perspective into account and to be aware of the fact that people in poverty are often reluctant to accept policy intervention, especially in the lives of their children. Successful early childhood programmes generally have in common the strong and consensual participation of parents. The growing interest of science and policy in this life stage may hold promise for the future of the EU and its youngest citizens.

REFERENCES

BLACKWELL, D.L., HAYWARD M.D., CRIMMINS, E.M. (2001), Does childhood health affect chronic morbidity in later life?, *Social Science & Medicine*, (52): 8, 1269–1284.

BOURDIEU, P. (1986), The forms of capital, in: J. RICHARDSON (ed.), *Handbook of Theory and Research for the Sociology of Education.* (New York: Greenwood), 241–258.

BROOKS-GUNN, J. & DUNCAN, G. (1997), The effects of poverty on children, *Children and Poverty,* (7): 2 – Summer/Fall 1997.

BROOKS-GUNN, J., DUNCAN, G.J., KLEBANOV, et al. (1993), Do neighborhoods influence child and adolescent development?, *American Journal of Sociology,* (99): 2, 353–395.

CHASE-LANSDALE, P.L., GORDON, R.A., BROOKS-GUNN, J., et al. (1997), Neighborhood and family influences on the intellectual and behavioral competence of preschool and early school-age children, in: BROOKS-GUNN, J., DUNCAN, G.J., ABER, J.L. (eds.), *Neighborhood poverty: Vol.1 Context and consequences for children.* New York: Russell Sage Foundation, 79–118.

DUNCAN, G.J., KALIL, A., ZIOL-GUEST, K. (2008), *Economic costs of early childhood poverty.* Issue paper no. 4. Partnership for America's economic success.

DUNCAN, G. & MAGNUSON, K. (2010), Promoting the healthy development of young children, in: Sawhill, I. (ed.), *One percent for kids.* Washington, D.C.: Brookings, 16–39.

EVANS, G.W. (2004), The environment of childhood poverty, *American Psychologist,* (59): 2, Feb-Mar 2004, 77–92.

ESPING-ANDERSEN, G. (2004), Untying the Gordian knot of social inheritance, *Research in Social Stratification and Mobility,* (21): 115–138.

FORS, S., LENNARTSON, C., LUNDBERG, O. (2009), Childhood living conditions, socioeconomic position in adulthood, and cognition later in life: exploring the associations, *Journal of Gerontology: Social Sciences,* 64B(6): 750–757.

KESTILA, L., KOSKINEN, S., MARTELIN, T., et al. (2005), Determinants of health in early adulthood: what is the role of parental education, childhood adversities and own education?, *European Journal of Public Health,* (16): 3, 305–314.

KNUDSEN, E., HECKMAN, J., CAMERON, J., et al. (2006), Economic, neurobiological, and behavioral perspectives on building America's future workforce, *Proceedings of the National Academy of Sciences,* (103): 10155–10162.

KRAMER, M.S. (2000), Socio-economic disparities in pregnancy outcome: why do the poor fare so poorly?, *Paediatric Perinatal Epidemiology,* (14): 3, 194–210.

LARSON, P.C., (2007), Poverty during pregnancy: Its effects on child health outcomes, *Paediatric Child Health,* (12): 8, 673–677.

LUO, Y. & WAITE L.J. (2005), The impact of childhood and adult SES on physical, mental, and cognitive well-being in later life, *Journal of Gerontology,* 60B: S93–S101

MACKENBACH, J., MEERDING, W., KUNST, A. (2007), *Economic implications of socio-economic inequalities in health in the European Union.* Brussels: European Communities.

SE´GUIN, L., XU, Q., GAUVIN, L., et al. (2005), Understanding the dimensions of socioeconomic status that influence toddlers' health: unique impact of lack of money for basic needs in Quebec's birth cohort, *Journal of Epidemiological Community Health,* 2005;59:42–48.

SHAKLEE, H., & FLETCHER, J. (2002), Key studies that rocked the cradle: How research changed the way we care for infants and toddlers, in: STEWART, B., LOVINGOOD, R., PURCELL, R. (eds.), *Research applications in family and consumer sciences*. Alexandria, VA: American Association of Family and Consumer Sciences.

SHONKOFF, J., & PHILLIPS D. (2000), *From neurons to neighborhoods: the science of early childhood development*. Washington, DC: National Academy Press.

SPC (2008), *Child Poverty and Well-Being in the EU. Current status and way forward*. European Commission. Directorate-General for Employment, Social Affairs and Equal Opportunities.

WADSWORTH MEJ & KUH D (1997), Childhood influences on adult health: a review of recent work from the British 1946 national birth cohort study, MRC National Survey of Health and Development, *Paediatric and Perinatal Epidemiology*, (11): 2–20.

WALDFOGEL, J. (2004), *Social mobility, life chances and the early years*. CASEpaper 88. ISBN 1460-5023.

WHELAN, C. & MAÎTRE, B. (2009), Comparing poverty indicators in an enlarged European Union, *European Sociological Review Advance Access*, 23 October 2009; doi: doi:10.1093/esr/jcp047.

WORLD HEALTH ORGANISATION (2007), *The World Health Report 2007 – A safer future: Global public health security in the 21st century*. Geneva: WHO.

VLAAMS AGENTSCHAP ZORG & GEZONDHEID (2007), *Geboorte en bevalling*. Retrieved 25/09/2007, from www.zorg-en-gezondheid.be/geboorte.aspx

10. A SCHOOL IN THE NEIGHBOURHOOD, A NEIGHBOURHOOD IN THE SCHOOL

Isabelle PANNECOUCKE

INTRODUCTION

Despite developments in and changed views on education and learning, school life still takes place mainly within the school walls, behind a gate keeping the outside world at a safe distance. The first section of this chapter shows that this kind of separation is not absolute and the school environment is not a neutral given. We illustrate this in the second section by looking at research focusing on children living in three urban neighbourhoods in Flanders: Antwerp North, Old Borgerhout and Merksem. We conclude the chapter with some policy recommendations.

1. THE CONTEXTUALISED SCHOOL

The shapes or designs of school buildings illustrate the fact that schools are not isolated but embedded in a broader societal context. The design of a school reflects a certain vision of education and directs some of its educational processes. According to Leemans (2006: 91), the increasing differentiation of classrooms and other rooms in the school building is related to the growing rationalisation and scientification of educational processes. Other than rationalisation, disciplining processes also play a role in the specific design of schools.

The location of the school building is another element that plays its role in daily school life. This certainly applies to primary schools, which are more often situated closer to children's homes than secondary schools (Creten et al., 2000). That way, the chances are high that schools in poor neighbourhoods mainly attract pupils from poor families. This homogeneous composition of pupils can affect their academic achievement and risks perpetuating and aggravating the problems arising from poverty and deprivation.

The surrounding neighbourhood surpasses the context of the school, as is illustrated by the existence of so-called concentration schools (Van Den Driessche, 2007). The specific socio-economic and sociocultural characteristics of the neighbourhood affect all aspects of the school. Leemans (2006: 73) states that the quality of school infrastructure is not an isolated given, but is part of the broader societal context. In his research in Brussels on Flemish education he establishes that infrastructural difficulties at the schools are related to the poverty problem in the school environment. Lower quality school infrastructure means an additional burden for those educational institutions already facing specific poverty-related problems.

Children from deprived neighbourhoods represent an additional challenge for these institutions. They come to school bringing their own specific background, which is often not well known by teachers and does not fit in a priori with the educational system. They walk through the school gate carrying a backpack filled with other capabilities. A different kind of social and cultural capital, an unstable home environment and health problems oppose the education system's ability to work as a lever for an escape from poverty (Pannecoucke, 2005).

1.1. EXPERIENCING SCHOOL LIFE

Because of this interdependence between neighbourhood and school it is important to pay attention to the spatial context and to the way children experience school life. Verschelden (2002: 11), amongst others, indicates that experience of school life is related not only to the real situation at school, but also to the specific family situation, the social and cultural context in which children live and the location of the school in the broader societal context. The feelings evoked by school life are illustrated by the description below of the subjective attitudes of children towards their schools. We focus on three urban neighbourhoods in Antwerp (Pannecoucke, s.d.).

1.2. A MEETING PLACE

When the children are asked what they think of their schools, the answers are predominantly positive. They perceive their schools as nice and fun. It is not so much learning different things that satisfies them; their daily conversations with their peers too are concentrated within the school's social dimension (who is in their class, which friends they have at school). We find – as do De Groof and Stevens (2004: 45) – that attending classes is not always what fascinates them. It is being able to meet their peers which makes going to school pleasing. Verschelden (2002: 311) and Van Gils (1992) come to a similar conclusion. The school is for children obviously more than an educational institution. It is also a meeting place, which is a latent function of the educational system (Laermans et al., 2001).

When we compare the three selected urban neighbourhoods, we discover that this kind of meeting place function is relatively more important to the children of Old Borgerhout than to the others. Contact with class peers is not limited to school hours. After school they meet at the mosque, while shopping (especially for girls), or when playing on neighbourhood squares (for boys). Even though children from Merksem tell us they appreciate the contact with peers, they do not meet them outside school hours. Fleur (11 years old, of Belgian origin) calls the world inside the school walls *"the only place where I meet the others (classmates)"*. This does not mean that they do not meet other children after school. On the contrary, precisely because they do meet other peers at the music lesson, at the gym or at the youth club they do not feel the need (and take the time) to meet their classmates after school. Although they meet children other than their classmates, they are "congenial". This neighbourhood, just like Old Borgerhout, can be defined as an "archipelago society": they grow up in Belgian society but each within their own specific networks. Bouw and Karsten (2004: 196–197), among others, find that people search for other people who are like-minded in a diverse world. The internet is another way for children from Merksem to maintain social contacts.

In Antwerp North the children are also satisfied with being able to make contacts in school. As in Merksem, these encounters are limited to school hours not so much because they meet other children in leisure activities but because of the lack of meeting space, resulting from the specific characteristics of the neighbourhood. We describe this group of children as "indoor children" who – because of the more disadvantaged financial situation at home – have fewer alternatives in leisure time.

The meeting function of the school is made extra clear in Old Borgerhout for girls. The school organises dance classes at a city youth centre in the neighbourhood and basketball games in cooperation with other schools.

> *On Wednesday after school I had to go dancing. Today we had a recital. We had to show our dances to a big audience. I had to dance in the Branderij: there is a big room. I danced with girl friends of my class at school. There were many girls.* (Fragment from the diary of Latifa, 11 years old, of Moroccan origin).

Through these activities these girls not only develop social skills and physical capacities; they are also given room to expand their action radius. Because they have to help in the household and have less freedom to walk around in the neighbourhood, they face a relatively limited use of space. In addition, they are able to make use of the internet in the youth centre, which is also a facility they often have to miss at home.

> *I went dancing in the Branderij at 13.30h with a couple of classmates. After dance, I used the PC. With the same classmates. I used MSN* (Fragment from the diary of Nadia, 12 years old, of Moroccan origin).

Other than being a neighbourhood facility that makes encounters between peers possible, the school also opens up access to other neighbourhood services, especially for girls.

1.3. A PLACE TO PLAY

A remarkable given with the children of Old Borgerhout is their appreciation of the school playground. It offers them the space to play, which is something not available to them at home and in their neighbourhood. In contrast to the children in Merksem, these children do not live in big houses with a garden in which they can play. In the neighborhood of Old Borgerhout there are fewer and smaller green playgrounds and sports fields to play soccer and other games. Moreover, girls also experience more limitations on going outdoors in the neighbourhood.

In Antwerp North, children are confronted with the same limitations as in Old Borgerhout: fewer green areas, less playground space and less freedom to play outdoors. This last element can be ascribed to personal characteristics and limitations imposed by parents. In this case, the limited opportunities to play in the neighbourhood are, however, not compensated by the availability of a big school playground. The playground is limited to a small walled area and the school is also relatively small, as Loubna and Sanaa point out. The "play" aspect of the school as a neighbourhood facility is for them less significant than for the children of Old Borgerhout.

> *The walls are so small, it is like a prison.* (Loubna, 12 years old, of Moroccan origin)
> *It's a bit of a small school.* (Sanaa, 12 years old, of Turkish origin)

1.4. A PLACE TO LEARN

It is remarkable that the children of Antwerp North – compared to children from the other two cases – perceive the variety of things they learn at school more positively. Children from Merksem, for example, point out much more that they attend school because they are compelled to, by their parents or by law. The appreciation of learning new things in Antwerp North is related to the value the children attach to education as a whole. They go to school to learn, to obtain a diploma or to get a better life later on.

> *Interviewer: Why do you go to school?*
> *Luna (12 years old, of Nigerian origin): To learn and to get a good profession later on.*
> *Hamid (13 years old, of Moroccan origin): To get smarter and to get a good profession.*
> *Ikram (12 years old, of Turkish origin): To learn, and when I am big, I want to have a good and decent life.*

They assume that if they are committed to their school career, this will be rewarded later on. Even though few people in their neighbourhood have a diploma and a paid job, they attach importance to education for their future. This finding is contrary to the collective socialisation perspective (Jencks & Mayer, 1990; Gephart, 1997; Leventhal & Brooks-Gunn, 2000), which presupposes the contrary. This can be related to the fact that the socialisation model implies social interactions with neighbours, which is less the case in Antwerp North. Because they do not know their neighbours well, the chances are smaller that they will look upon them as role models. The children are, however, aware of the problems in their neighbourhood and of the problems some of the inhabitants have to deal with. This might stimulate them not to wind up in the same situation and to see education as a way out. Such an attitude is also translated into "higher" professional aspirations, such as wanting to be physicians and pharmacists.

2. CHOOSING A SCHOOL

In choosing a school for secondary education different factors play a role (Creten et al., 2000; Creten & Douterlungne, 2001; Karsten, 2002). The range of motives for school choice can be divided into two categories: educational and non-educational (Driessens et al., 2003). Educational arguments have to do with quality. According to Vogels (2004), this is a concept that is hard to define, because its operationalisation is subject to societal developments. In trying to clarify the concept, Vogels distinguishes the quality of the input (the school building, the personnel, teaching material), the quality of the process (didactic approach, educational climate) and the quality of the output (educational achievements).

Non-educational motives are especially related to the demographic characteristics of the parents and the students, the direction of the school, the availability of after-school activities and the advice of others (Driessens et al., 2003).

These general motives are, however, weighed against the actual availability of educational institutions in the vicinity. Creten et al. (2000) suggest that the motives induced by the local offering (the "local" motives of choice) actually lead to the eventual choice of school. This choice, in other words, takes place in relation to the actual offering.

Figure 1. Location of secondary schools in Old Borgerhout

Source: Pannecoucke, 2009.

Legend:
1. Second and third grade of technical education (trade) and vocational education (trade)
2. Second and third grade of general secondary education (economics), technical education (trade, fashion, nursing) and fourth grade of vocational education
3. Second and third grade technical education (nursing), vocational education (trade and nursing)
4. First grade vocational and general education
5. Second and third grade vocational education (carpentry)
6. Second and third grade general education

It is also clear from our research that distance plays an important role. Especially for children with a limited action radius, as is the case in Old Borgerhout, the element of proximity is important. Letting distance playing a role limits the number of potential schools and types of education and schooling (see figure 1). In the map, the secondary schools are indicated by a number. The legend describes the types of education pupils can gain at each school.

The nearest schools offer only technical and vocational education in the first and second grade of secondary school. In school number two, pupils can take the second and third grade of general secondary education, but this school is situated somewhat further away. School number six also offers general secondary education, but is located on the other side of the big road R1. The children experience this road as a geographical neighbourhood boundary, which results in a perception of this school as hard to reach.

The choice of one of the nearest secondary schools will hamper children's advancement into higher education. They often follow the lead of older brothers and sisters and of their classmates. It is clear that the segregation in primary school education is continued into secondary education.

A similar map is found in Antwerp North (see figure 2).

Figure 2. Location of secondary schools in Antwerp North

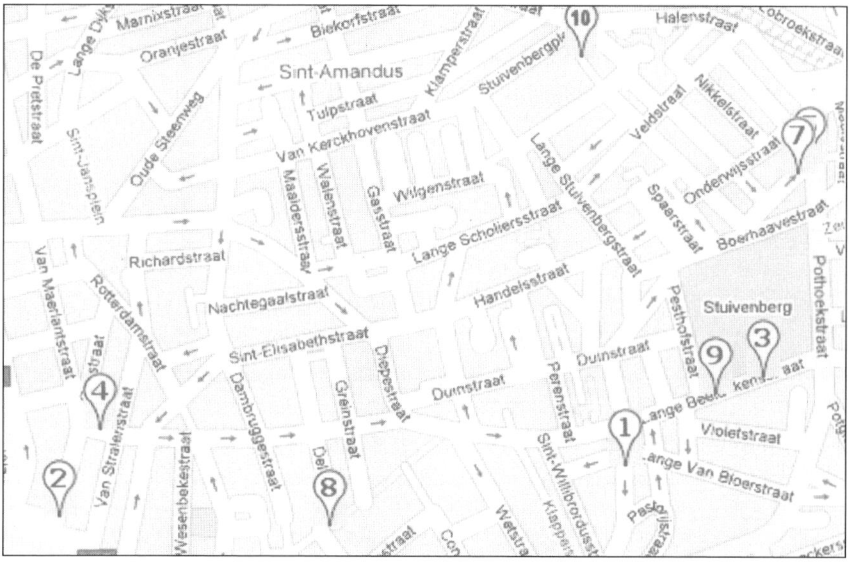

Source: Pannecoucke, 2009.

Legend:
1. Second and third grade of vocational, art, and technical education; fourth grade of vocational education
2. Second and third grade of general, technical and vocational education (plus class for newcomers)
3. Second and third grade of vocational education (nutrition) and technical education (chemistry, nursing)
4. First grade general, technical education
5 through 7. Second and third grade of vocational education (trade), technical education (trade)
8. Second and third grade of vocational education (fashion and nursing)
9. All types of education (limited choice of programmes)
10. Second and third grade of general and technical (trade) education

However, there is a different story in the case of Antwerp North, in which the teacher plays an important role. She takes the children on school visits to the other bank of the City (Left bank) and to another school, also out of the neighbourhood. Both schools are outside the neighbourhood and offer general secondary education, as opposed to the schools nearest to where the children live. The teacher takes them on this visit for two reasons. Firstly, she perceives the neighbourhood in which the children grow up as an area which offers few opportunities, amongst other factors because of the limited number of educational programmes provided by the schools. Secondly, she realises that the children have only known a limited use of space. By having these children experience just how far these schools are in reality and how they can get there, they gain a better view of the accessibility of the location. These school visits do have consequences: several children take the bus or tram to go to classes in one of the secondary schools offering general education. Take Brahim for example:

> ... *because transportation is good. And my friends go there.* (Brahim, 12 years old, of Turkish origin)

Contradicting the hypothesis of the stigmatisation perspective (Musterd & Gouthals, 1999; Forrest & Kearns, 2001), the negative perception of the teacher does not turn against the children. On the contrary, it motivates her to show the children the schools outside the neighbourhood.

For children from Merksem, the third case studied, distance is not a criterion in the choice of a secondary school. On the one hand, this is related to the relatively unlimited use of space. When they want to participate in leisure activities, neighbourhood boundaries do not stop them. They also have a large range of educational choice nearby (see figure 3).

Figure 3. Location of secondary schools in Merksem

Source: Pannecoucke, 2009.

Legend:
1. Second and third grade of general education
2. Second and third grade of general education
3. Second and third grade of general education, technical education (trade, nursing, tourism)
4. Second and third grade of general education, technical education (nursing)
5. Second and third grade of general education (general and sports education), technical education (sports)
6. Second and third grade of vocational education (body care), technical education (chemistry, body care)
7. Second and third grade of general education
8. Second and third grade of technical education (trade, societal security, sports), vocational education (carpentry, nutrition, societal security)
9. Second and third grade of vocational education (trade, nutrition), technical education (chemistry, optician, dental technician, nutrition)

Letting distance play a role in school choice obviously limits the number of potential schools and types of education. Children from Merksem, however, have a wide variety of schools and a diversified offering of types of education. Their luck somehow seems to be double: they have more options to choose a school and their wider action radius also gives them a larger number of potential schools and educational types to choose from.

3. LESSONS FOR POLICY MAKERS

For children, school is more than just a place where they learn things. It is also a place where they meet different peers. In Old Borgerhout the school also seems to open access to other neighbourhood organisations. The school is also a place to play. Among other things this is connected to the relatively limited action radius of girls and the lack of room to play at home and in the neighbourhood. Precisely because the school is located in the neighbourhood it is important to take this into account. Can the school stimulate participation in organised activities? Can the school broaden playing opportunities for the children?

Children also grow up and learn outside school. It is therefore important that the school is embedded in these contexts. This kind of "anchoring" can be achieved in two ways: by bringing the world of the school into the neighbourhood and by bringing the neighbourhood into the school. Which route is followed, depends on the context.

Being an actor in the local setting is a huge challenge both for educational institutions and for policy-makers. The need for organised activities and opportunities for play is not really apparent in the most affluent neighbourhoods. Schools which want to contribute to these functions are especially those with children from socio-economically disadvantaged families who also live in more deprived neighbourhoods. However, these schools' buildings, where such activities could be developed, are often of inferior quality.

The idea of the neighbourhood playing a role in the vicinity of schools is slowly seeping into policy making. There is, for example, a regulation giving priority to the enrolment of children from the neighbourhood. Next to brothers and sisters and children benefiting from priority rules (Equal Education Opportunities policy), neighbourhood children also have priority for enrolment in primary school. The local alderman responsible for education policy sees this as an opportunity to develop children campuses. This kind of campus can offer a wider range of child care, preschools and primary schools to the neighbourhood and is intended to provide an alternative for the children living in it. A neighbourhood school allows for small distances to have to be covered and for more local involvement in the school. The local alderman (Voorhamme, 2008) wants *"open schools with involvement of parents and the neighbourhood"*.

Recognition of the importance of the neighbourhood environment is also found in the adjusted financing of the school based on the characteristics of the school and its pupils. Next to the educational level of mothers, entitlements to a school or study allowance and the home languages, the neighbourhood of the school is an additional element in school financing.

The Flemish Minister of Education (Vandenbroucke, 2006) also states that broad living and learning environments, providing new opportunities for the children, can be created by responding to local opportunities and needs. An important policy focus is that these opportunities are unequally distributed. We therefore suggest that initiatives working on promoting equal opportunities, lifelong learning for deprived groups and meaningful leisure activity need to receive the necessary policy attention and funding (Nicaise et al., 2004: 100).

The conclusions on the choice of secondary school also provide some lessons for policy making. Distance to school appears not to be equally important to all children. We established that the accessibility of a school is a factor in school selection in Old Borgerhout, and for them the nearest schools only offer vocational and technical education in the second and third grade. Children from Merksem – who do not experience distance as a problem – actually have a wide variety of schools in the neighbourhood.

Delivering a more diversified offering of secondary schools and educational programmes in deprived neighbourhoods seems to be an appropriate policy recommendation in this respect. A more equal distribution of schools, however, would not solve all problems. The presence of a school in the neighbourhood providing general education does not automatically mean that children will go to school there, because some perceive general secondary education as "not for them". Nevertheless, factors such as distance and accessibility would not be as important as they are today.

A less drastic solution could be to respond to the motives for school choice. Our research results indicate that the teacher can mitigate the element of distance. If pupils were assisted to experience for themselves what their home-school itinerary could look like if they chose to attend a secondary school offering a broader range of educational opportunities further away from home, it would contribute to a growing awareness that certain schools are accessible although not located in the neighbourhood. In this research, this kind of teacher attitude is related to her perception of the neighbourhood in which she teaches, and her knowledge that the pupils have a limited action radius. Because teachers generally do not live in the neighbourhoods in which they teach, working towards an objective perception and knowledge of deprived neighbourhoods can inspire teachers to develop initiatives and activities, and to respond to the limitations and opportunities of the neighbourhood.

Finally, classmates also seem to play a role in the choice of a secondary school. Most of the children take the decisions of their classmates into account when making their own choice. The segregation found in primary schools is likely to be followed by segregation in secondary education. The consequences of segregation as such are more than only not learning to deal with the diversity in the society. According to Oberti (2007: 4), the school not only reproduces existing social and ethnic divides in urban space, but also enhances polarisation. The level of segregation in certain schools is much larger than is spatial segregation.

Despite efforts made under the Flemish "Equal Education Opportunities" Policy, segregation in Flemish schools remains a reality. The potential consequences of and for school population composition should be kept in mind in designing new policy measures.

REFERENCES

BOUW, C. & KARSTEN, L. (2004), *Stadskinderen. Verschillende generaties over de dagelijkse strijd om ruimte*, Amsterdam: Aksant.

CRETEN, H. et al. (2000), *Voor elk wat wils: schoolkeuze in het basis- en secundair onderwijs*, Leuven: KUL, HIVA.

CRETEN, H. & DOUTERLUNGNE, M. (2001), *Schoolkeuze van ouders en leerlingen in het basis- en secundair onderwijs in relatie tot opleidingsniveau en levensbeschouwing van de ouders*, (online), www.hiva.be/docs/paper/paper_hilde_creten.pdf, last accessed 1 April 2007.

DE GROOF, S. & STEVENS, F. (2004), Over onderwijs, in: Burssens, D. et al. (eds.), *Jeugdonderzoek belicht. Voorlopig syntheserapport van wetenschappelijk onderzoek naar Vlaamse kinderen en jongeren (2000–2004)*, Gent: K.U.Leuven, VUB & UGent, 31–58.

DRIESSEN, G. et al. (2003), *Sociale Integratie in het Primair Onderwijs*, Amsterdam/Nijmegen: SCO-Kohnstamm Instituut/ITS.

FORREST, R. & KEARNS, A. (2001), Social Cohesion, Social Capital and the Neighbourhood, *Urban Studies*, 38 (12): 2125–2143.

GEPHART, M. A. (1997), Neighborhoods and Communities as Contexts for Development, in: Brooks-Gunn, J., Duncan, G. J. & Aber, J. L. (eds.), *Neighborhood Poverty. Volume I. Context and Consequences for Children*, New York: Russell Sage Foundation, 1–43.

JENCKS, C. & MAYER, S. (1990), The social consequences of growing up in a poor neighbourhood, in: Lynn, L. E. & McGeary, M. F. H. (eds.), *Inner-city poverty in the United States*, Washington D.C.: National Academy Press, 111–186.

KARSTEN, S. et al. (2002), *Schoolkeuze in een multi-etnische samenleving*, Amsterdam: SCO-Kohnstamm Instituut van de Faculteit der Maatschappij- en Gedragswetenschappen.
LAERMANS, R., VANHOVE, T. & SMEYERS, M. (2001), *Beeldvorming en beschrijving van de leefwereld van jongeren*, Leuven: K.U.Leuven, Centrum voor cultuursociologie.
LEEMANS, G. (2006), Schoolgebouwen in relatie tot onderwijs en samenleving, *Tijdschrift voor onderwijsrecht en onderwijsbeleid*, 2005-2006 (1-2): 85-96.
LEVENTHAL, T. & BROOKS-GUNN, J. (2000), The Neighborhoods They Live in: The Effects of Neighborhood Residence on Child and Adolescent Outcomes, *Psychological Bulletin*, 126 (2): 309-337.
MUSTERD, S. & GOETHALS, A. (1999), *De invloed van de buurt*, Amsterdam: SISWO.
NICAISE, I., PIRARD, F. & L. REULENS (2004), Naar een brede school in Vlaanderen, *Impuls*, 35(2): 99-108.
OBERTI, M. (2007), Social and school differentiation in urban space: inequalities and local configurations, *Environment and Planning A*, 39 (1): 208-227.
PANNECOUCKE, I. (s.d.), *Buurt en School: Gescheiden Contexten? Een onderzoek bij kinderen naar de relatie tussen hun buurt en hun schoolervaringen en aspiraties*, te verschijnen.
PANNECOUCKE, I. (2005), Kansarmoede en secundair onderwijs: kwantitatieve en kwalitatieve probleemsituering voor Vlaanderen, *Tijdschrift voor onderwijsrecht en onderwijsbeleid*, (1-2): 20-33.
PANNECOUCKE, I. (2009), De school in de buurt, de buurt in de school, in: VRANKEN, J. et al. (eds.), *Armoede en Sociale Uitsluiting. Jaarboek 2009*, Leuven: Acco, 181-192.
VANDENBROUCKE, F. (2006), *Brede school in Vlaanderen en Brussel*, toespraak 11 december 2006, Brussel, Hendrik Consciencegebouw
VAN DEN DRIESSCHE, M. (2007), De tocht van de kinderen, *Tijdschrift voor Architectuur OASE*, 72: 72-97.
VAN GILS, J. (1992), *Wie niet weg is, is gezien. Hoe beleeft het kind zijn gezin, zijn school en zijn vrije tijd?*, Brussel Koning Boudewijnstichting.
VERSCHELDEN, G. (2002), *Opvattingen over "welzijn" en "begeleiding". Een sociaal-(ped)agogische analyse van leerlingenbegeleiding als exemplarisch thema in het jeugdbeleid*, Gent: Academia Press.
VOGELS, R. (2004), *Ouders bij de les. Betrokkenheid van ouders bij de school van hun kind*, Den Haag: Sociaal Cultureel Planbureau.
VOORHAMME, R. (2008), *Buurtkinderen kunnen voorrang krijgen*, (online), www.robertvoorhamme.be/2008/buurtkinderen-kunnen-voorrang-krijgen.html, gelezen op 09-10-2009.

11. LIKE A CHILD'S GAME: A POLICY CONFIGURATION APPROACH TO CHILD POVERTY

Danielle Dierckx

INTRODUCTION

In this final chapter a framework for the analysis of child poverty policies is presented. Child poverty is a complex topic that challenges public policies. We focus on public policy mechanisms and processes and confront these with the special requirements a policy theme like child poverty eradication raises for governments. We therefore use the policy configuration approach developed in earlier poverty policy research (Dierckx, 2007).

1. THE POLICY CONFIGURATION APPROACH

The policy configuration approach allows us to make a holistic description and analysis of current child poverty policies. Furthermore, it establishes the features, possible merits, the threats and opportunities, and conditions for so-called inclusive and interactive policy-making.

This approach has been developed in order to get a grip on complex and multidimensional policy issues, like poverty (Dierckx, 2007). It answers questions such as: What are the elements we have to take into account if we want to describe, analyse, evaluate and adjust a particular policy? Central to the framework is the concept "policy configuration", meaning the totality of the constituting parts of policy-making and of policy principles, measures, structures and actors involved. The policy configuration approach consists of the interplay between three main components: the policy theme, the policy organisation and the policy style. In order to undertake the study of a policy configuration, we need to take account of those three components and their mutual linkages. This may be visualised as follows (Figure 1).

Figure 1. Analytical framework of the policy configuration approach

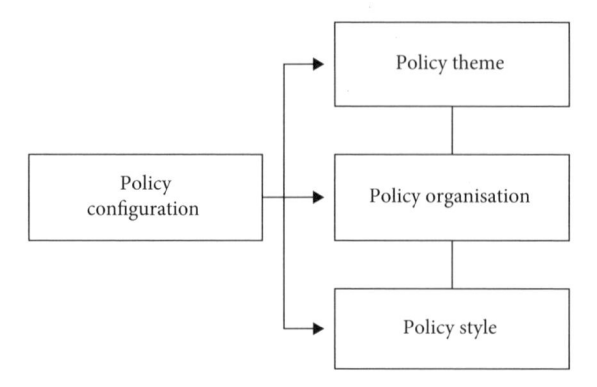

```
                                    ┌─────────────────┐
                              ┌────▶│  Policy theme   │
                              │     └─────────────────┘
                              │              │
┌──────────────────┐          │     ┌─────────────────┐
│     Policy       │──────────┼────▶│ Policy organisation │
│  configuration   │          │     └─────────────────┘
└──────────────────┘          │              │
                              │     ┌─────────────────┐
                              └────▶│  Policy style   │
                                    └─────────────────┘
```

Source: Dierckx, 2007.

The realisation of inclusive and interactive policies requires conditions to be fulfilled for each of those components. For each component we will discuss some theoretical and empirical insights related to child poverty.

2. CHILD POVERTY AS A POLICY THEME

What is the use of analysing child poverty as a policy theme? The Thomas theorem (1928) states it clearly: "If men define situations as real, they are real in their consequences". The way a problem is defined and accepted by policy-makers determines the policy actions that will be undertaken. In this section we focus on the mechanisms for defining child poverty and strategies for tackling it.

2.1. A BUNCH OF DEFINITIONS

Different actors have different views on the same (part of) reality, in this case child poverty, which is illustrated by the presence of more than one definition of child poverty. Each definition is shaped by a particular world view and by particular interests (also called the "self-referential view"; Van Twist in Lips, 2001). This has implications for the way solutions are designed and implemented, such as through public policies.

Several contributors to this book emphasise this diversity of perspectives and definitions. We could organise some of those differences according to the following dichotomies:

- *income* poverty versus multidimensional poverty or *deprivation*
- *material* versus *immaterial* poverty

- children in poverty as a *homogeneous* or *heterogeneous* group
- children *in* poverty versus *child poverty*
- *structural* versus *contextual* approaches
- *universal* versus *context-specific* concept of child poverty.

Different perspectives become visible when strategies are formulated on how to tackle child poverty. Again, different actors have their own opinions, driven by their own interests. This is illustrated in Morrow's contribution, where she states that policy-makers are using a dominant economic vision of childhood. She formulates it as a challenge to discard this vision, because it stresses only the importance of the future utility of children in the economic dimension of daily life. It represents a human capital approach, based on "outcomes" and realising children's potential as productive, hard-working adults. She pleads for a shift towards a better understanding of what childhood is like now and a corresponding change in public policies and practices.

Seemingly opposite is the statement Roelen makes when formulating a dichotomy of well-being versus well-becoming. She discovers a lack of long-term perspective, in particular in the underlying discourse on the way in which child poverty is measured. She states that the concept of well-being reflects children's well-being in the present without taking their future potentials into account.

Vandenhole warns against the negative effects of introducing a rights-based approach (RBA) in policy discourse and strategies. The RBA has the underlying assumption of the autonomy of the individual, he says, and therefore bears the risk of individuating and decontextualising poverty. It is not unlike the traditional liberal presupposition that everyone is born equal and should be treated as such in her/his life, including when clearly in a situation of social or economic inequality – thus increasing this inequality.

Different opinions may lead to policy controversies (Schön & Rein, 1994). In the end, most statements in this book refer to the importance of striving for a holistic and inclusive perspective on children in a context of poverty.

2.2. THE LIE OF GOOD INTENTIONS

In a policy process, defining child poverty as a policy theme and formulating goals to tackle it is followed by the development and implementation of specific policy measures. Analysing policies means to recognise the (potential) existence of several "reality levels". At the surface there is policy practice; underneath we find more general policy goals and declarations of intention; at the deepest level, discourse is established. In this stage of policy analysis, one should be aware of the possible contrasts between the policy measures to combat child poverty and the formal discourse of government and non-governmental actors. In other words, to what extent does a discrepancy exist between words and deeds?

This statement is illustrated by the findings of Vandenhole. The state's responsibility with regard to poverty is a subsidiary one, he says, which is, moreover, centred on material conditions and access to social services. Whether this approach corresponds to the current understanding of the nature of poverty should be questioned. Vandenhole formulates it as a challenge concerning turning the conceptual advantages of the rights-based approach into practice. Other authors, such as De Boyser and Vranken, also stress the difference between a broad, multidimensional definition of child poverty and the limited indicators that are being used to underpin public policies.

2.3. AGENDA SETTING

That nowadays child poverty has reached the European policy agenda and several national ones is due to a combination of factors. Traditionally, poverty is confronted by a lot of prejudice and creating the necessary public and political support is not easy. The fact that poverty is being linked to a specific target group or category, in this case children, makes a difference. As Vandenhole writes: "The broadly shared assumption that children are in a very vulnerable situation makes them the prototype of "deserving poor", i.e. those that deserve to be assisted and protected".

Table 1. Social construction of categories for public policies

	Positive label ("deserving")	Negative label ("non-deserving")
Much power	1. Advantaged E.g. firms, middle class, academics	2. Contenders E.g. feminists, trade unions, holebis
Little power	3. Dependants E.g. lone parent families, youth, disabled people ("deserving poor")	4. Deviants E.g. criminals, drug addicts ("non-deserving poor")

Source: based on Schneider & Ingram (1997; 2003).

We refer to the social construction approach of Schneider and Ingram (1997; 2003) to clarify the mechanisms behind this phenomenon. They point out that governments are more in favour of some target groups than others and therefore are more or less willing to undertake supportive policy measures. Schneider and Ingram distinguish four categories in terms of the way they are socially constructed by public authorities (positive/negative) and of the extent of their political power.

The "advantaged" are powerful and have a positive label. The "contenders" are also powerful but are categorised as "undeserving" or greedy, obviously negative labels. The "dependants" have a positive label; this category is good but relatively needy or it concerns people who are considered helpless with little or no power. The "deviants" are also powerless, but they bear the stigma of being "undeserving", violent and mean (cf. also Gans, 1995).

The categories in this typology are not homogeneous. For example, people experiencing poverty might be divided into deserving and non-deserving poor and thus might be categorised as respectively "dependants" and "deviants". Depending on the category, public policy measures differ. The "deviants" are not entitled to support; on the contrary, they need to be punished. The "dependants" deserve support, because they bear no responsibility for their situation. Categories might change position in this frame. Mostly, a change happens because of economic, social or political circumstances or of changing perceptions.

Children are mainly perceived as dependent, with a positive label but without much power. Both features are stressed where children experiencing poverty are concerned. That they have to live in poverty is not in the first place seen as their responsibility. Their position within the typology may change in two ways. Children may lose their positive label if, for example, their behaviour, norms and values are perceived as deviant, such as when they disturb public safety or public order. Policy measures then become less supportive and more punitive. Illustrations are youth prisons and obligatory placements in special child and youth care by the court. Children may also make a shift towards the category of the advantaged. This means that they keep their positive label but enhance their power. Initiatives that empower children and enable them to participate in decision-making processes theoretically contribute to this shift. Another strategy for reaching a more powerful position is by building coalitions and by strengthening social networks through the inclusion of more powerful groups. An illustration is the installation of a children's rights officer or efforts by labour unions to improve child protection, abolish child labour and increase support for one-parent families.

Roose et al. warn of the risk of restricting policy efforts to children and so isolating them from the larger context in which they are living. "A contextualised approach to children's rights has an eye for the possible negative consequences of some interpretations of children's rights, such as a diminishing understanding of the position of parents in poverty and a loss of solidarity between children and adults."

3. A HOLISTIC APPROACH GUARANTEED BY GOVERNANCE STRUCTURES

Tackling child poverty implies that several policy fields come into view, such as education, health, culture and leisure time. The daily life of children is automatically coloured by these domains and requires policies to have certain features. Government has to organise itself in such a way that it covers the policy fields involved and the connections between those fields. Vandenhole adds: "Inserting the (H)RBA (rights-based approach) principles into policies and programmes also narrows policy discretion, in that a minimum threshold is established; process requirements such as participation and prioritisation of the poor, and of poor children in particular, are introduced; and the best interests of the children are taken into account." So the specificity of child poverty considered within a rights-based approach has its impact on the way government can act.

When translated into policy practice, this statement means that governments must establish a formal structure that realises and guarantees policy attention and policies on this matter. Child poverty should be reflected in the so-called policy organisation (or the "institutional design", see Goodin, 1996; Hall & Taylor, 1996; Lowndes, 2001). A multidimensional policy issue such as child poverty is assumed to put lots of pressure on the government's (internal) organisation. Mainstreaming, cross-cutting policies and inclusive policies express the aspiration to develop a holistic approach (Perri, 1997). This approach requires government to make choices on how to structure and combine a territorial, sectoral and categorical approach to the reality which it wants to change. How should multi-level governance be organised? Which sectors and policy fields should be involved? What categories of people should be defined as target groups? Most of all, the question arises how to coordinate these different perspectives.

How should a policy organisation that guarantees coordination be operationalised? An attempt is made in figure 2, where a coherent set of tools is presented.

Figure 2. Policy tools for an inclusive and coordinated policy

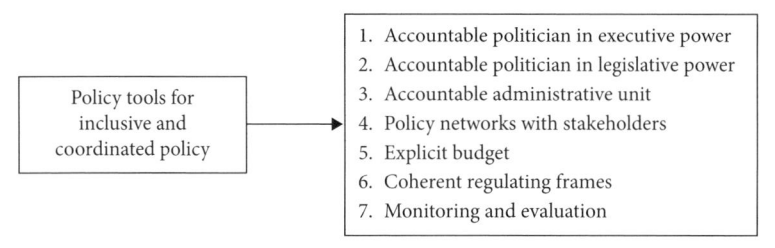

Policy tools for inclusive and coordinated policy →
1. Accountable politician in executive power
2. Accountable politician in legislative power
3. Accountable administrative unit
4. Policy networks with stakeholders
5. Explicit budget
6. Coherent regulating frames
7. Monitoring and evaluation

Source: Redig & Dierckx, 2003.

The tools concern the different organisational and institutional aspects of a specific policy configuration. We formulate a brief outline with those tools, point by point.

Ad 1. Responsible and accountable politicians in executive power
- Politicians in executive power are e.g. ministers and aldermen;
- Responsibility and accountability have a double meaning, namely to be responsible for and to be formally addressable (accountable) about the policy domain concerned;
- Responsibility and accountability can be shaped concretely by including the policy domain explicitly in the title of the political office or, in any case, by formally and explicitly assigning this authority;
- As the visibility of the policy concerned increases, the policy actions gain in power.

Ad 2. Responsible and accountable units in legislative power
- Units with legislative power are, for example, parliamentary commissions and city councils.
- Here the same definitions and criteria concerning responsibility and accountability are applicable;
- The surplus value of increasing the visibility of the policy domain within legislative institutions is connected to the space and opportunities for (more) thorough debate in order to stimulate evaluation and reorientation…

Ad 3. Responsible and accountable administrative units
- These are municipal, provincial and national administrative cells, which are specifically authorised for that policy domain;
- The respective administrations must also develop a clearly demonstrable and recognisable identity;
- It is important to have an eye for the unit's hierarchical position within the entire administration. The higher it is placed, the more competently and powerfully it can function.

Ad 4. Policy networks with stakeholders
- Within government, formal policy networks ensure that there is enough communication and cooperation between the different parties involved (Carlsson, 2000; Marsh & Rhodes, 1992);
- Networking involves constructing crossroads, both within the same government level (horizontally) and between the different levels (vertically – local, provincial, national). Those networks build bridges between legislative and executive power, with the involvement of administrations;
- It also includes the main stakeholders from civil society and representatives from the target groups or policy categories.

Ad 5. Explicit budgets
- Policy intentions and plans remain idle words if the essential financial resources are not allocated to them. An investigation of the budget is an inseparable part of every policy evaluation;
- To obtain a complete picture of the whole of the resources reserved, it is usually necessary to collect and bundle the specific and often diffuse elements of policy;
- This exercise makes it possible to match the budget and to monitor it in the near and distant future.

Ad 6. Explicit and coherent regulating frames
- The regulating frames are the entirety of laws, decrees, ordinances and other rules concerning the policy domain in view;
- To track possible disintegration, these should be screened for coherence, corresponding frame of concepts and complementary objectives;
- An instrument that prevents different policy options and measures from neutralising or opposing each other could be a (child/youth) impact assessment.

Ad 7. Scientific research: monitoring and evaluation
The policy is pursued on the basis of current information about the policy subject. Different quantitative and qualitative analyses will increase the quality of the policy and predict future challenges. The handling of scientific measuring instruments keeps the permanent evaluation of the pursued and planned policy in the picture. The researcher is an ally when it comes to putting the policy theme on the political agenda.

4. WHO PLAYS A DECISIVE ROLE?

The third and last component of a policy configuration is the policy style, which focuses on what kind of democracy is implemented. Does government keep all the activity and responsibility in policy-making to itself ("delegation to government") and/or does it allow other relevant non-governmental actors to participate in policy-making ("citizen participation")? Policies to combat child poverty may be conducted from different perspectives on how policies should be developed. In this book the rights-based approach provides some indications, such as stating the importance of child participation. Within the policy configuration approach this participation is considered to be a way to deliver information that inspires policies and policy-making. In this final section we pay attention to two kinds of information delivery. The first concerns information from the children experiencing poverty; we call it the life experience approach. The second focuses on the role of academics, or the evidence-based approach.

4.1. A LIFE EXPERIENCE APPROACH

In many contributions in this book the participation of children is promoted. Vandenhole refers to Gallagher to caution against over-optimism. The question needs to be asked to what extent children have the capability to resist power and even change power relations through participation. Power is used to persuade "people to participate in their own subjection" but it also "equips (a human being) to become an independent actor" (Gallagher, 2008). In his view, "participation has the potential for both compliance and insubordination. One may wonder whether and to what extent poverty impinges on children's ability not only to resist power, but also to change power relations and the concomitant social exclusion."

In addition, Morrow refers to article 12 of the UN Convention on the Rights of the Child. In this article participation is referred to as one of the four general principles of the Convention. It stipulates that "[s]tate parties shall assure to the child who is capable of forming his or her own views of the right to express those views freely in all matters affecting the child, the views of the child being given due weight in accordance with the age and maturity of the child…". She also warns against too much optimism and suggests as one of the alternative strategies that research can be developed that is sensitive to the views of children and respects it; Pannecouckes's research provides us with a nice illustration of an effort thus inspired.

Roose et al. argue that participation is not an inherently positive and unambiguous notion because it mirrors ideological motives of social policy, civil society and practice. This is illustrated by the distinction between participation as an instrument and as a point of departure. The first approach contains the risk that clients of social work practice will be excluded when they fail to show willingness to participate as actively as social workers and policy-makers expect of them. In the latter approach, the rights of children and parents are seen as being shaped through engagement in a participative process in which the definition and content of these rights are negotiated.

Apart from these critiques, it is obvious that the perspectives of children may enrich policies, at least in principle. They may contribute to a better understanding of the lives of children experiencing poverty. A more accurate problem definition should improve the efficiency of policies. It should be clear that these targets can only be reached with tailor-made participation methods.

4.2. AN ACADEMIC EVIDENCE-BASED APPROACH

Another strategy for supporting and informing policies for combating poverty is to collect evidence through academic research. Although qualitative research results in interesting in-depth information (see e.g. Pannecoucke's contribution), the majority of research is quantitative, such as (child) poverty measurement.

The way child poverty is defined determines the kind of indicators that are used. Roelen states that a lack of a clear understanding of the meaning of the term child poverty can lead to inadequate or inappropriate use of the child poverty measure to hand. That is why in her contribution De Boyser uses a multiple deprivation measure, next to the at-risk poverty rate.

Quality criteria for a good measurement method include avoiding unnecessary complexity, measuring material and social deprivation, basing poverty lines on social norms, establishing a monitoring system, and building public support for poverty reduction.

5. CONCLUSION

What are the key success factors for a sustainable and efficient child poverty policy as viewed from the policy configuration approach?

Based on the considerations above, we would emphasise the importance of a match between the formal discourse (articulation of the view on what has to be done) and the underlying ideas, goals and intentions in policy measures. In order to strengthen the policy discourse, a broad consensus amongst governmental and non-governmental actors involved in child and/or poverty matters should be realised. To put child poverty eradication high on the policy agenda it is useful to create public and political support. Another key success factor is the presence of advocates: people occupying powerful positions in society and/or politics and arguing in favour of developing a child poverty policy.

Governments need to organise themselves better to deal with the ambitions of a coordinated, inclusive approach, which means that policy-makers should be made accountable for its development. If no policy-maker bears clear responsibility, no one will make an effort to establish sustainable and convincing policies. Traditionally, policy competences are divided by sector (such as employment, housing, education) and rarely by categories. An institutional design that intends to tackle child poverty effectively implies policy networks that cut across multiple sectors and that take account of the specific requirements of a children's perspective.

The evidence from children's life experience may complement the classical evidence of scientific research. Governments therefore need to create opportunities and space for more interactive policy-making. Investing in appropriate participation and research methods is the key to realising information input and measurement that ensure enduring respect for the theoretical insights developed.

REFERENCES

CARLSSON, L. (2000), Policy networks as collective action, *Policy Studies Journal*, (28): 502–521.

DIERCKX, D. (2007), *Tussen armoedebeleid en beleidsarmoede. Een retrospectieve en interventiegerichte analyse van de Vlaamse beleidspraktijk* (*Between poverty policy and policy poverty. A retrospective and intervention oriented analysis of Flemish policy practice*), Leuven/Belgium: Acco.

GANS, H. (1995), *The war against the poor. The underclass and antipoverty policy.* New York: Basic Books.

GALLAGHER, M. (2008), Foucault, Power and Participation, *International Journal of Children's Rights*, (16): 3, 395–406.

GOODIN, R.E. (1996), *The theory of institutional design.* Cambridge: Cambridge University Press.

HALL, P.A. & Taylor, R.C.R. (1996), Political science and the three new institutionalisms, *Political Studies*, (XLIV): 936–957.

LIPS, M. (2001), De beleidstheorie vanuit diverse wetenschaps- en beleidspraktijken: van 'hoe verder' naar 'hoe anders'?, in: Abma, T. & In 't Veld, R. (eds.), *Handboek beleidswetenschap*. Amsterdam: Boom, 271-280.

LOWNDES, V. (2001), Rescuing Aunt Sally: taking institutional theory seriously in urban politics, *Urban Studies*, (38): 11, 1953-1971.

MARSH, D. & RHODES, R.A.W. (1992), *Policy networks in British government*. Oxford: Clarendon Press.

PERRI 6 (1997), *Holistic government*. London: Demos.

REDIG, G. & DIERCKX, D. (2003), Armoedebeleid van overheden: over het aanleggen van kruispunten en rotondes (Public anti-poverty policies: developing crossroads for an integrated policy), in: Vranken, J., De Boyser, K. & Dierckx, D. (eds.), *Poverty and social exclusion. Yearbook 2003*. Leuven/Leusden: Acco, 335-354 (published in Dutch) (www.oases.be).

SCHNEIDER, A.L. & INGRAM, H. (1997), *Policy design for democracy*. Studies in government and public policy, Lawrence, Kansas: University Press of Kansas.

SCHNEIDER, A. & INGRAM, H. (2003), Social construction of target populations: implications for politics and policy, *The American Political Science Review*, (87): 2 (June 1993), 334-347.

SCHÖN, D.A. & REIN, M. (1994), *Frame reflection. Towards the resolution of intractable policy controversies*. New York: Basic Books.

VAN TWIST, M.J.W. (1993), De beleidstheorie vanuit de wetenschapspraktijk: Van 'hoe ver?' naar 'hoe verder'?, *Beleidswetenschap*, (1): 34-74.

ABOUT THE AUTHORS

Nina Biehal is research director of the Social Work with Children and Young People team in the Social Policy Research Unit at the University of York.

Maria Bouverne-De Bie is head of the Department of Social Welfare Studies at the University of Ghent, Belgium.

Katrien De Boyser is senior researcher at the Centre for Research and Development Monitoring (at the University of Ghent, research group Macro & Structural Sociology), and free collaborator at the Research Centre on Poverty, Social Exclusion and the City (University of Antwerp). Her main research interests lie in social mobility and (early) childhood poverty.

Danielle Dierckx is lecturer and post-doctoral researcher at the Department of Sociology at the University of Antwerp. She is director of the Centre on Inequalities, Poverty, Social exclusion and the City (OASeS). For more information, see www.oases.be or www.ua.ac.be/danielle.dierckx.

Virginia Morrow is reader in Childhood Studies at the Institute of Education, University of London, where she is programme leader of the MA Sociology of Childhood and Children's Rights. Children and young people have been the focus of her research activities since 1988. Her main research interests are the history and sociology of child labour and children's work; methods and ethics of social research with children; sociology of childhood and children's rights; social capital in relation to children and young people; children's understandings of family and other social environments. She is an editor of *Childhood: a global journal of child research*. She is the author of numerous papers and reports.

Ides Nicaise has a background in economics and works as a research manager at HIVA (Research institute for Work and Society – Catholic University of Leuven). He further specialised in social policy, more precisely the relationships between education, labour market policy and social inclusion (in rich as well as in developing countries). He is the Belgian member of the EC's independent expert group on social inclusion. Besides his professional activities, he is chairing the Belgian Resource Centre for the Fight against Poverty, a centre created by law as an interface between the government, the civil society and grassroots organisations defending the interests of the poor.

About the authors

Isabelle Pannecoucke is a postdoctoral researcher at the Center for Social Theory (University of Ghent). She defended her Ph.D. dissertation 'Buurt en school: gescheiden contexten?' (Neighbourhood and school: separated contexts?) in 2009. In this dissertation, she has studied the relationship between the neighbourhood of children, their school experiences and their perceived life chances. Currently, her research focuses on the dynamics of gentrification processes in city neighbourhoods.

Gwyther Rees is research director for the national charity The Children's society, in the United Kingdom.

Didier Reynaert is research assistant at the Department of Social Work and Welfare Studies at the University College Ghent, Belgium.

Keetie Roelen works as a researcher and project consultant for Maastricht Graduate School of Governance at Maastricht University in the Netherlands. She holds a MSc. in International Economics and PhD degree on the subject of child poverty measurement and policy. Keetie Roelen's research interests include poverty, poverty reduction policies and social protection. She has worked as a consultant on a number of training and research projects for international organizations, including the World Bank, UNICEF and UNDP in Europe, Asia and Africa, including Moldova, Kosovo, Namibia and Vietnam.

Griet Roets is assistant professor at the Department of Social Welfare Studies, Ghent University, Belgium.

Rudi Roose is assistant professor at the Department of Social Welfare Studies, Ghent University and associate professor at the Department of Criminology, Free University Brussels, Belgium.

Ingrid Schockaert is a PhD student in Demography at the University of Louvain-la-Neuve. She has worked on issues of labour market participation and fertility, policy evaluation and migration. Currently she is a member of HIVA's research group 'Poverty, social inclusion and migration'.

Wouter Vandenhole teaches human rights and holds the UNICEF Chair in Children's Rights at the Faculty of Law of the University of Antwerp (Belgium). He is the co-director of the Law and Development Research Group, University of Antwerp Law Research School. His research interests include economic, social and cultural rights, children's rights, and the relationship between human rights law and development. UNICEF respects the academic freedom of the chair holder. Opinions expressed by the chair holder do not commit UNICEF.

Jan Vranken is emeritus full professor of the University of Antwerp (Belgium). He founded the Centre OASeS (Centre on Inequality, Poverty, Social Exclusion and the City), in which he still is very much engaged; the same goes for his participation in local, national and international initiatives outside academia. For more information: www.ua.ac.be/jan.vranken.